高等职业教育数控技术专业规划教材
国家示范性高职院校建设项目成果

现代数控机床原理与结构

（中英双语）

郁元正　编
鲁聪达　审

机械工业出版社

本书以国际化制造业人才培养为目标，通过中英文结合的方式介绍了当代常用数控机床的结构与原理。在内容的编排上，深入浅出，突出数控技术的先进性和实用性。本书以英语为主，中文作为概括及解释，详略视内容的难易程度而定。文中配有大量插图，以利提高记忆效果，并减少对中文注释的依赖。对于来自国外资料的英语原文作了少量改动，减少长句和生词量，增强了可读性。

全书共分为七章，分别介绍了数控机床的发展简史、数控机床的基本传动方式与坐标系、数控机床的典型零部件，以及数控车床、数控铣床、加工中心、电加工机床等四类常见数控机床的工作原理与应用。

本书配有电子课件，凡使用本书作教材的教师可登录机械工业出版社教材服务网（http://www.cmpedu.com）下载，或发送电子邮件至cmpgaozhi@sina.com索取。咨询电话：010-88379375。

本书可以作为开展双语教学的高等院校使用，也可作为数控技术、机电一体化技术等专业的专业英语教材。为了改善学习效果，建议读者在使用本教材之前对数控机床具备基本的知识和操作经历。

图书在版编目（CIP）数据

现代数控机床原理与结构：（中英双语）/郁元正编．—北京：机械工业出版社，2013.8

高等职业教育数控技术专业规划教材

ISBN 978-7-111-42403-1

Ⅰ.①现… Ⅱ.①郁… Ⅲ.①数控机床—理论—双语教学—高等学校—教材—汉、英②数控机床—结构—双语教学—高等学校—教材—汉、英 Ⅳ.①TG659

中国版本图书馆CIP数据核字（2013）第092732号

机械工业出版社（北京市百万庄大街22号 邮政编码100037）

策划编辑：郑 丹 王英杰 责任编辑：王英杰 王晓艳 武 晋

版式设计：霍永明 责任校对：纪 敬

封面设计：鞠 杨 责任印制：李 洋

北京华正印刷有限公司印刷

2013年7月第1版第1次印刷

184mm×260mm·11.25印张·275千字

0001—2000册

标准书号：ISBN 978-7-111-42403-1

定价：23.00元

前　言

教育要面向现代化、面向世界、面向未来。近年来，制造业的国际技术合作、贸易日益增多，国外设备大量引进。无论使用、维护、研究进口设备与产品，还是与国外同行进行顺利的交流，扎实的专业知识和必要的英语技能缺一不可。因此，英语能力逐渐成为高技能专业人员的必备素质，也是扩大信息来源、进一步学习国外先进技术的基本条件。

为了适应机械加工行业的蓬勃发展和教育改革的不断深化，我院于2006年启动了数控技术专业中澳合作班的双语教学工作。经过多年的教学实践和不断地总结经验，深刻认识到配套双语教材的重要性。然而，目前各校的双语教学尚处在摸索探讨阶段，未形成统一的模式，教材也各有千秋。为了体现教改内涵，使双语教学更好地开展，以国家示范性高职院校建设为契机，在学院领导和教务处的支持下，以中澳合作办学项目为载体，编写了这本《现代数控机床原理与结构（中英双语）》。

本书吸收了近几年出版的数控机床教材和数控技术专业英语教材的优点，同时具有以下特点：

（1）全书以英文表述为主，配有大量插图，图文并茂，直观易懂，以降低阅读难度。

（2）每一单元末尾列出了本单元词汇，按字母顺序列表，以便读者查询。

（3）在保持该学科知识体系完整的基础上，循序渐进，分散难点，着重应用。

因为可以借鉴的资料不多，加之编者初次尝试，水平有限，不妥之处和缺点疏漏敬请各位读者批评指正。

编　者

Contents

Chapter 1 Introduction

1.1 Background

1.1.1 The development of CNC machine tools

After the World War II, with the increasing complexity of aircraft components, such as turbine blades of jet engines, in 1949, the US air force asked Parsons Company to develop a special equipment to automatically machine those components.

With the concept of numerically controlled (NC) machine tool by Parsons Company, in 1952, a control device that based on the vacuum tube (see Figure 1-1) and relay has been developed for the milling machine control. It is known as the first generation of NC machine tool. Figure 1-2 shows a turbine blade of jet engine.

数控机床的发展历程——

第二次世界大战后，美国空军为了实现日益复杂的飞机零部件的自动加工而委托 Parsons 公司研制一种机床数字控制系统。

Figure 1-1 Vacuum tube（电子管）

Figure 1-2 A turbine blade of jet engine

By the end of 1950s', the second generation of the NC machine tool whose control circuits mainly consisted of transistors (see Figure 1-3) was successfully developed. With the development of NC machine tools, practicability, flexibility, convenient maintenance and adaptability were required by users. Figure 1-4 shows a control cabinet.

20 世纪 50 年代末，数控机床控制电路中的电子管（图 1-2）逐步被晶体管（图 1-3）取代。图 1-4 所示为控制柜。

However, transistor circuits cannot meet all these requirements due to economy. More over, large volume of the circuits requires a big control cabinet. In 1965, with the development of integrated circuits (see Figure 1-5), the third generation of NC machines provides solutions for those requirements.

1965 年，集成电路（图 1-5）应用于机床控制，进一步提高了控制系统的性价比。

Figure 1-3 Transistor

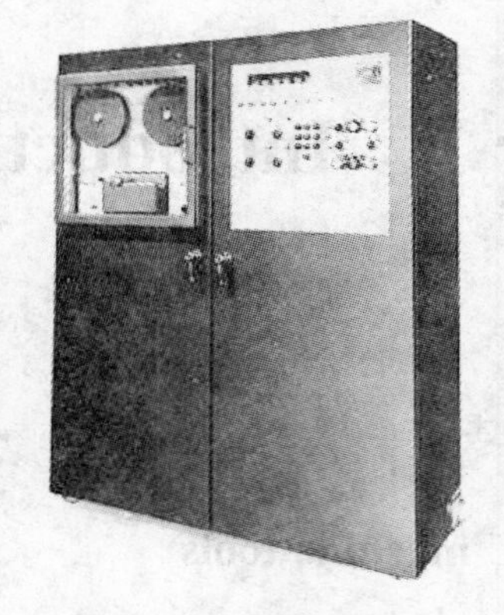
Figure 1-4 Control cabinet

Figure 1-5 An integrated circuit

Applying integrated circuit on machine tool controller has experienced three development stages:

➢ Stage1 1950s-1970s Numerical control (NC)

➢ Stage2 1970s-1980s Computer numerical control (CNC). Microprocessors significantly improved characteristics and reliability of control systems. From then on, CNC technology was rapidly developed and widely applied all over the world.

➢ Stage3 1990s-present PC platform.

1.1.2 Definition of CNC machine tools

CNC technology: Controlling mechanical movement and working process by numerical signals.

CNC machine tools (see Figure 1-6): Machine tools controlled by the programable system

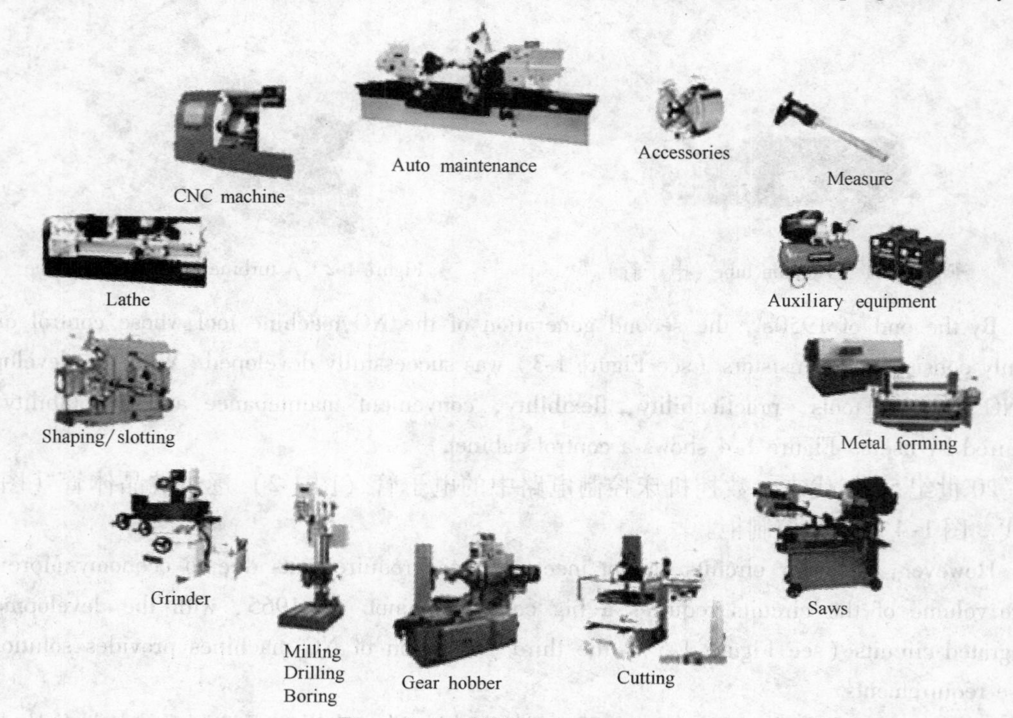

Figure 1-6 CNC machine tools

that can logically execute special codes.

数控技术的定义：用数字信号对机械运动和工作过程进行控制的技术。

数控机床（图 1-6）的定义：一种装有程序控制系统的机床，该系统能逻辑地处理具有特定代码、编码指令的程序。

When machining with a conventional machine tool, operators manipulate hand wheels to move the cutting tool along the desired contour of work piece. In a CNC machine tool, the CNC system controls all motions according to the program, as shown in Figure 1-7.

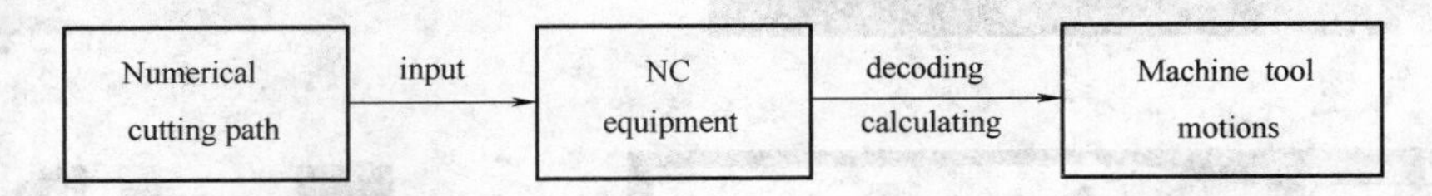

Figure 1-7 Basic work principle of CNC machine tools

1.2 Construction and working principle of CNC machine tools

1.2.1 Construction of CNC machine tools

CNC machine tools have different purposes in various fields, however, any CNC machine tool has a similar construction, namely, the control medium, CNC device, servo mechanism, assistant control system and machine tool reality, as shown in Figure 1-8.

数控机床的种类很多，但任何一种数控机床都是由控制介质、数控装置、伺服系统、辅助控制系统和机床本体等若干基本部分组成的，如图 1-8 所示。

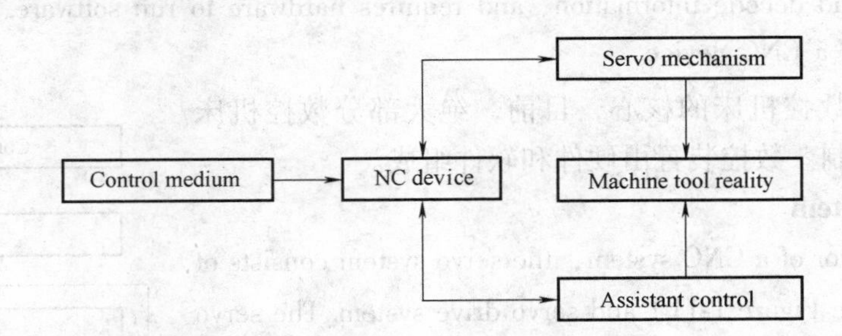

Figure 1-8 Basic construction of a CNC machine tool

1. Control medium

The CNC machine tool requires no direct manual operation during working, but must perform as the operator required. A "bridge" between human being and the machine tool is known as the control medium (see Figure 1-9) (also known as the program carrier), carrying all required machining information for the CNC device.

Typically, the control medium includes punched tape, punched card, magnetic tapes, floppy disks and flash disks, which depends on the CNC device. Machining information must be transferred to the CNC device through readers, such as photoelectrical tape readers, tape machines, disk drivers and USB ports.

控制介质（也称为程序载体）承载着所有的加工信息（程序），如图 1-9 所示，它们是人的思想与机床运动之间的桥梁。这些信息由专门的阅读设备读取。

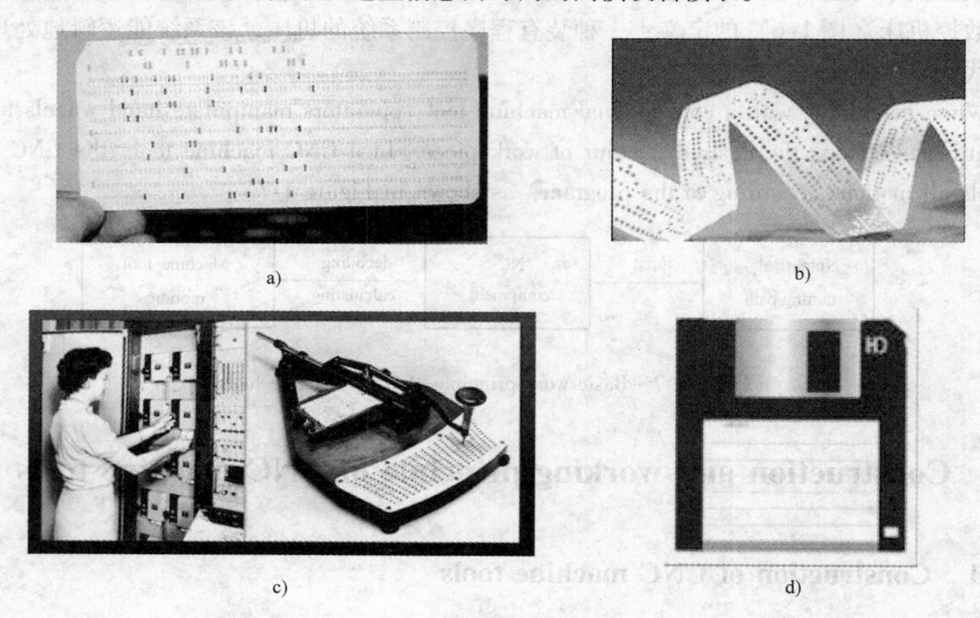

Figure 1-9　Control media

a) Punched card　b) Punched tape　c) Tape puncher　d) Floppy disk

2. CNC control equipment

Today, most CNC machining tools are controlled by microprocessors. The CNC device requires software to read and decode information, and requires hardware to run software. Figure 1-10 shows the construction of a CNC device.

数控装置是数控机床的核心，目前，绝大部分数控机床采用微处理器控制。数控装置由硬件和软件组成。

3. Servo system

As the executor of a CNC system, the servo system consists of a servo motor (see Figure 1-11) and servo drive system. The servo system drives the cutting tools or worktable, and/or actuates accessories to perform machining as required. The operating information is in the form of pulses. Each pulse produces a displacement of relevant movable part, which is known as pulse equivalent, usually equals 0. 001mm.

伺服系统由伺服驱动电动机（图 1-11）和伺服驱动装置组成，是数控系统的执行部分。数控系统控制伺服系统的指令信息是以脉冲形式体现的，每个脉冲使机床移动部件产生的位移量叫做脉冲当量。目前所使用的数控系统脉冲当量多为 0. 001mm。

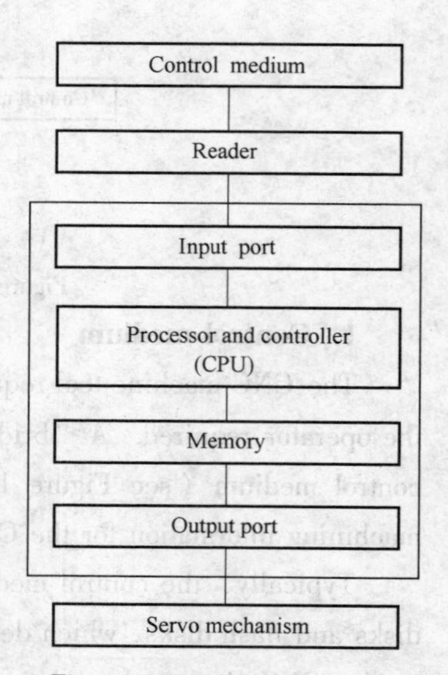

Figure 1-10　Construction of a CNC device

4. Assistant control system

The assistant control system is a sort of electric control equipments connecting the CNC device and mechanical components. It receives instruction signals (e. g. spindle speed, tool changing, and coolant supply) from the CNC device. After decoding, logic judgment and power amplification for the signals, the machine tool will be acting as desired driven by its electric, hydraulic, pneumatic or mechanical appliances. The oil pump (see Figure 1-12) is usually an important component of an assistant control system. Moreover, assistant control systems sometimes produce switching signals to CNC.

辅助控制系统可根据指令信号控制主轴转速、刀具更换、切削液供给等。

Figure 1-11 Servo motor

Figure 1-12 Oil pump

5. Machine tool reality

The machine tool reality is the main body of a CNC machine tool, consisting of a basis (e. g. base and bed) and movable components (e. g. worktable, carriage, spindle). Similarly, a CNC machine tool has all the mechanical components to perform machining processes.

For a CNC machine tool, it features of:

1) Driven by the high performance spindle and servo transmissions.

2) Better rigidity and anti-vibration of the structure.

3) Efficient transmission components, e. g. ball screw-nut pairs, rolling ways, etc.

Figure 1-13 shows the construction diagram of a conventional lathe.

机床本体是数控机床的主体，由基础部分（如基础、床身）和可动部件（如工作台、床鞍、主轴）组成。与传统机床类似，数控机床本体也是由执行机械加工的机械部件构成的。为了获得更好的切削性能和加工精度，与普通机床相比，各部件又有所改进。

1.2.2 Work principle of CNC machine tools

Generally, a CNC machine tool follows these steps to machine a part:

1) Programming according to drawings.

2) NC device reading machining program.

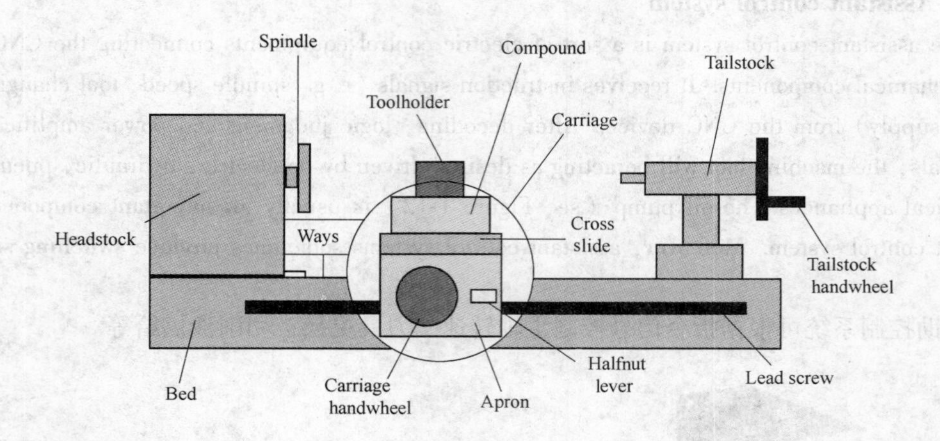

Figure 1-13　Construction diagram of a conventional lathe

3）Program decoding.

4）Servo mechanisms controlling machining process（spindle，feeding，tool changing，clamping，cooling，lubricating，etc.）

Figure 1-14 shows the work principle of a CNC machine tool.

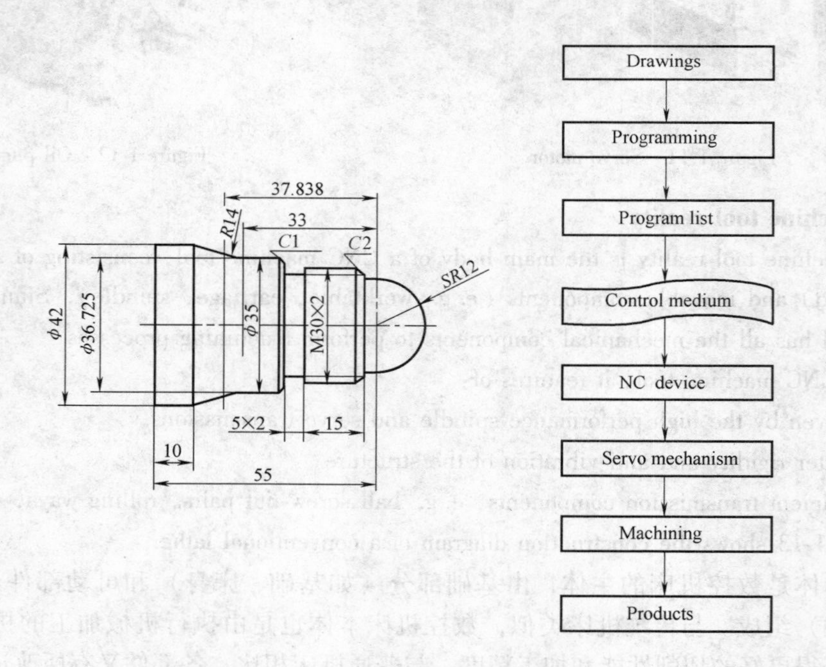

Figure 1-14　Work principle of a CNC machine tool

图 1-14 所示为数控机床加工过程示意图。按照零件图的技术要求和工艺要求，编写零件的加工程序，数控装置严格按照加工程序规定的顺序、轨迹和参数，通过伺服系统控制机床的各种运动，从而加工出符合图样要求的零件。

1.3 Classifications and features of CNC machine tools

1.3.1 Classifications of CNC machine tools

1. By functions（按工艺用途分类）

（1）General CNC machine tools Similar to conventional machine tools, the CNC lathe, CNC milling machine, CNC borer, CNC driller, CNC grinder, etc. were developed for different machining purposes. Each of them can be further classified, e. g. vertical milling machine, horizontal milling machine, gantry milling machine（龙门铣床）, etc.

They have similar functions as conventional machine tools, but can perform machining for parts with much higher accuracy and complex contour.

根据不同的工艺要求，数控机床也可分为车床、铣床、镗床、钻床及磨床等，其工艺可行性和通用机床相似，不同的是它们能加工精度更高、形状更复杂的零件。

（2）Machining center A machining center（see Figure 1-15）refers to a CNC machine tool with a magazine and cutting tool changing mechanism. Milling centers and turning centers are typical machining centers.

The machining center has better accuracy, efficiency, automation and lower running cost due to concentrated machining processes, therefore:

1）Reducing the number of machine tools required.

2）Reducing preparing time（e. g. tool alignment）before machining.

3）Reducing error between clampings.

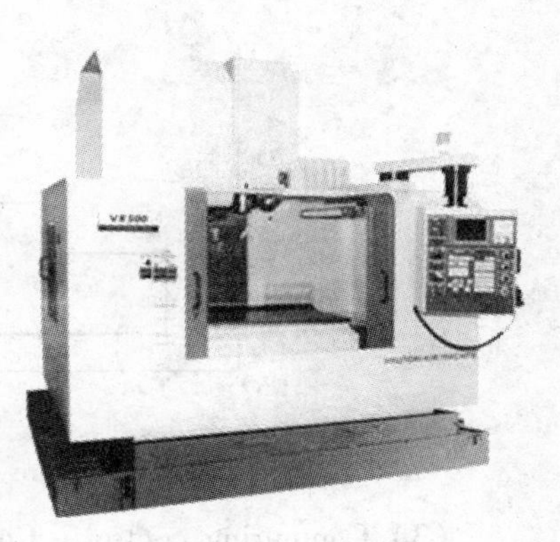

Figure 1-15 Machining center

数控加工中心是带有刀库和自动换刀装置的数控机床。在数控加工中心，零件一次装夹定位后，可进行多种工艺、多道工序的集中连续加工。与一般的数控机床相比，加工中心能实现更高精度、更高效率、更高程度自动化及更低的平均加工成本，因此近年来得到了迅速发展。

2. By motion types（按控制的运动轨迹分类）

（1）Point-to-point For this type, only the end point of motion need to be precisely controlled, i. e. precise locating from a point to another. The path and speed of the motion are not accurately controlled, and there is no machining process during moving and locating（see Figure 1-16）.

To reduce the time for moving and locating, moving parts usually rapidly approach the end point, and then precisely reach the target point at a low speed to ensure accuracy. Point-to-point control is applied on the CNC driller, CNC punch, CNC spot welding machine, etc.

点位控制的特点是只控制移动部件由一个位置到另一个位置的精确定位，对运动过程中的轨迹和速度没有严格要求，在移动和定位过程中不进行任何加工，如图1-16所示。

（2）Straight line control Besides the precise location of the start point and end point, straight

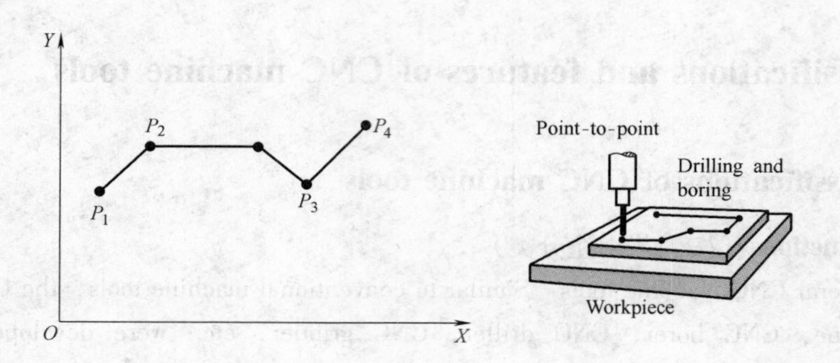

Figure 1-16　Motion contrail of point-to-point control

line control must ensure the moving path and speed (for different feedrate) between them. Machining is performed during moving (see Figure 1-17).

Usually, the cutting paths are parallel to coordinate axes, and sometimes inclined straight lines. CNC lathes, CNC grinders and CNC milling machines are typically under straight line control.

直线控制不仅要控制刀具相对于工件运动的两点间的准确位置，还要控制这两点间移动的速度和轨迹，如图 1-17 所示。

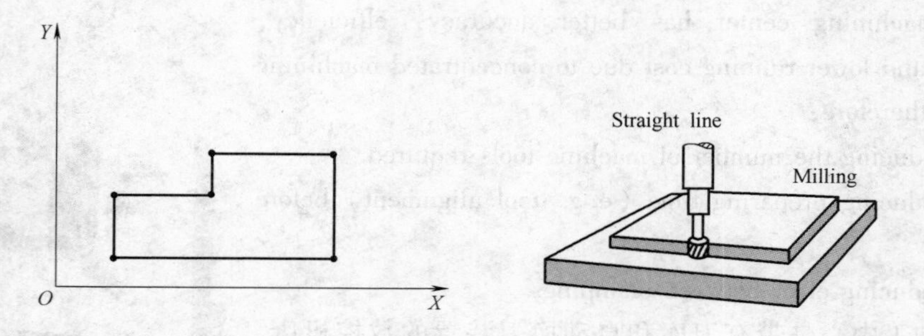

Figure 1-17　Motion contrail of straight line control

(3) Contouring control　Contouring control is also known as continuous control, which is applied on most CNC machine tools. It controls at least two axes at the same time, and performs interpolation (see Figure 1-18). Besides the start point, end point and moving path, moving speed at each point along the path needs to be controlled (see Figure 1-19).

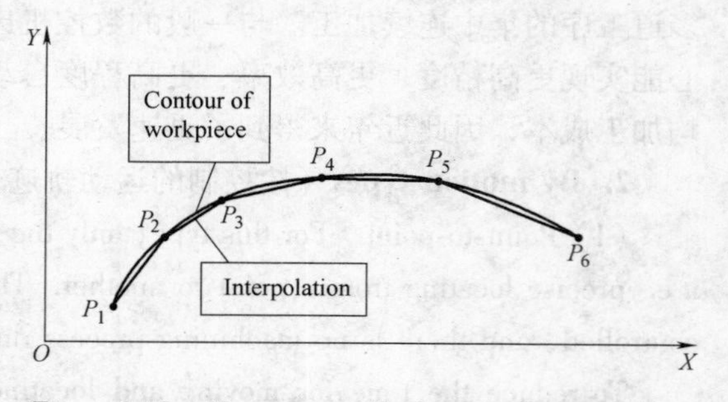

Figure 1-18　Motion contrail of contouring control

Under contouring control, parts with complex curves and curve surfaces can be machined. The control type is usually applied on the CNC lathe, CNC milling machine and machining center.

轮廓控制又称为连续控制，其特点是能同时控制两个以上的坐标轴，不仅要控制起点和终点位置，而且要控制加工过程中每个点的位置和速度，加工出由任意形状的曲线或曲面组

成的复杂零件，如图 1-19 所示。

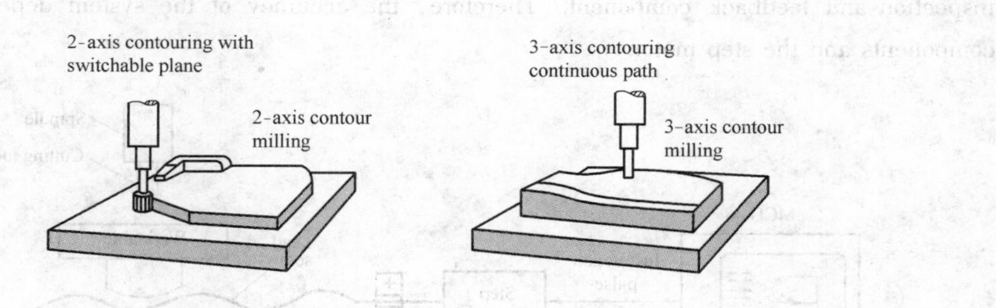

Figure 1-19 Motion contrails of 2-axis and 3-axis contouring

Thinking:

Could you please define the motion control types of machining processes in Figure 1-20?

Figure 1-20 Machining processes

a) Drilling b) Boring c) 3D milling d) Turning

3. By servo mechanisms（按伺服系统的类型分类）

According to inspection and feedback components, servo system on machine tools can be classified into:

① Open loop servo.

② Closed loop servo.

③ Semi-closed loop servo.

(1) Open loop control (see Figure 1-21) Under open loop control, the moving speed and

displacement of the worktable is determined by frequency and the number of pulse since no inspection and feedback component. Therefore, the accuracy of the system depends on driving components and the step motor.

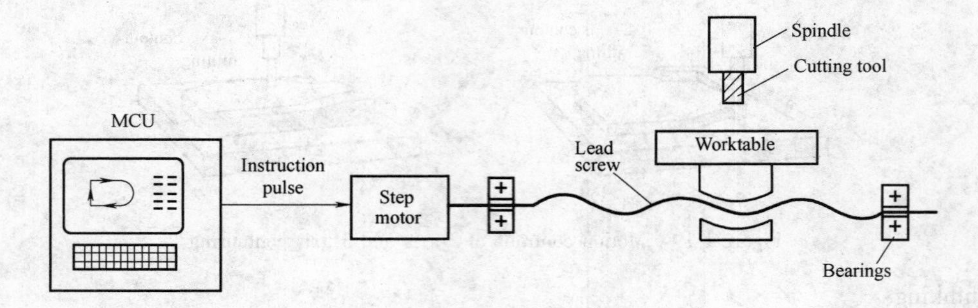

Figure 1-21 Open loop control

Open loop machine tools have following advantages:

1) Simple construction.

2) Low cost.

3) Convenient adjustment.

4) Stable performance.

Open loop control system is suitable for comparatively low accuracy applications, e. g. economical machine tools.

开环控制数控机床中没有检测反馈装置，其精度主要取决于驱动部件和步进电动机的精度和性能，如图 1-21 所示。这类数控机床结构简单、成本低、调试方便、工作比较稳定，适用于精度、速度要求不高的经济型、中小型机床。

（2）Closed loop control（see Figure 1-22） In this type, inspection components were built in worktable. CNC device compares instruction to real location of the worktable, and then corrects errors. Closed loop system can eliminate the effects of transmission errors. However, in fact, mechanical components have non-linear parameters, e. g. friction, rigidity, clearance, etc. It made parameter adjustment difficult. Any defect may result in system instability.

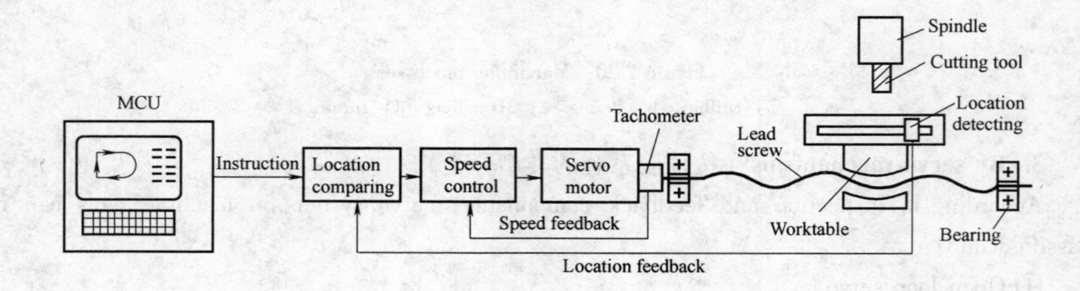

Figure 1-22 Closed loop control

Closed loop control system is usually applied on large machine tools where high precision is

required, e. g. CNC milling/boring machine, super-precision lathe, super-precision grinder, etc.

闭环控制数控机床的检测元件装在工作台上，数控装置根据指令信号与工作台末端测得的实际位置反馈信号的差值不断修正运动误差，可以得到很高的加工精度，如图 1-22 所示。但是，闭环控制系统的设计和调整都有较大的难度，因此主要用于一些精度和速度要求都较高的精密大型数控机床，如数控镗铣床、超精车床、超精磨床等。

(3) Semi-closed loop control (see Figure 1-23) It is applied on most CNC machine tools. For this type, inspection components are fixed at the end of the motor shaft. With inspection components that have high resolution, the system is able to provide reasonable accuracy, which is usually between that of open loop and closed loop. Moreover, the system is stable because of short control circuit.

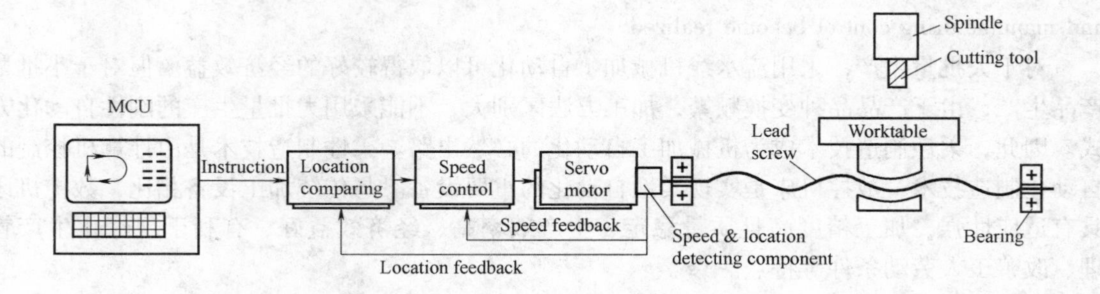

Figure 1-23 Semi-closed loop control

大多数数控机床都采用半闭环控制系统，它的检测元件装在电动机的轴端，由于传动链短，因此控制特性稳定，如图 1-23 所示。其控制精度介于开环系统与闭环系统之间。

1.3.2 Features of CNC machine tools

Flexible machining technology is known as computer controlled automatic production, including single CNC machining equipment and accessories. Therefore, CNC machine tools are the most important part for flexible automation. Compared to conventional machining facilities, CNC machine tools have following advantages:

1. Good adaptability

CNC machine tools are especially suitable for machining parts with complex contour and in small quantity.

2. High accuracy and stable production quality

Pulse equivalent of 0.001mm, high rigidity, thermal stability and clearance eliminating mechanism ensure high machining accuracy. Besides, CNC machine tools automatically perform machining thus avoid manual errors.

3. High production efficiency

Great rigidity allows large feedrate and cutting depth. Automatic changing machining position and cutting tools eliminates transporting, clamping and measuring between machining procedures.

4. Reducing manual labor and improving working environment

Automatic machining requires little manual operation and provides workers a better work environment, because workers are not necessary to stay in the workshop all the time during machining.

5. Cost-efficient

Though expensive capital cost increases depreciation（折旧），for small quantity production, CNC machine tools save time and cost for preparing, fixture, adjustment, and inspection. Stable and long term accuracy of machine tools reduces wasters. CNC machine tools can perform many tasks to save total number of machine tools.

6. Providing possibility for modern production management

CNC machine tools transfer and process information by standard codes, therefore, CAD, CAM and manufacturing control become realized.

对于大批量生产，采用流水线机械加工自动化可以取得较好的经济效益。但对于小批量产品生产，由于产品品种变换频繁，加工方法区别大，不能采用大批量生产的刚性自动化方式。因此，柔性制造技术成为机械加工自动化的必然出路。柔性制造技术是由计算机控制的自动化制造技术，数控机床是实现柔性自动化的重要设备。与传统加工设备相比，数控机床具有适应性强、加工精度高且质量稳定、生产效率高、经济效益好、有利于现代化生产管理、改善工人劳动条件等特点。

1.4 Future development

1.4.1 Development tendencies of CNC machine tools

① High speed, high efficiency, high accuracy and high reliability.

② Modulization, intelligentization, flexiblization and integration（模块化、智能化、柔性化和集成化）.

③ Users oriented.

④ A new generation of CNC machining process and equipments.

1.4.2 Modern manufacturing system

With the rapid development of flexible automation, single CNC machine tools are gradually replaced by machining center, FMC, FMS and CIMS.

(1) Flexible manufacturing cell (FMC) An FMC consists of 1-2 machining centers, industrial robots, CNC machine tools and material storage/transfer facilities. It has the flexibility for machining frequently updating products. FMC has been widely applied in industry.

柔性制造单元（FMC）由 1 ~2 台加工中心、工业机器人、数控机床及物料运储设备构成，具有加工多品种产品的灵活性，目前已进入了普及应用阶段。

(2) Flexible manufacturing system (FMS) An FMS is an automatic machining system consisting of CNC machining facilities, material storage/transfer facilities, and computer control

system (see Figure 1-24).

An FMS contains several FMCs. It can be rapidly adjusted for different machining tasks.

柔性制造系统（FMS）是由数控加工设备、物料运储设备和计算机控制系统组成的自动化制造系统，包括多个柔性制造单元，能迅速进行调整以适应频繁变化的生产任务，适用于多品种、中小批量生产，如图 1-24 所示。

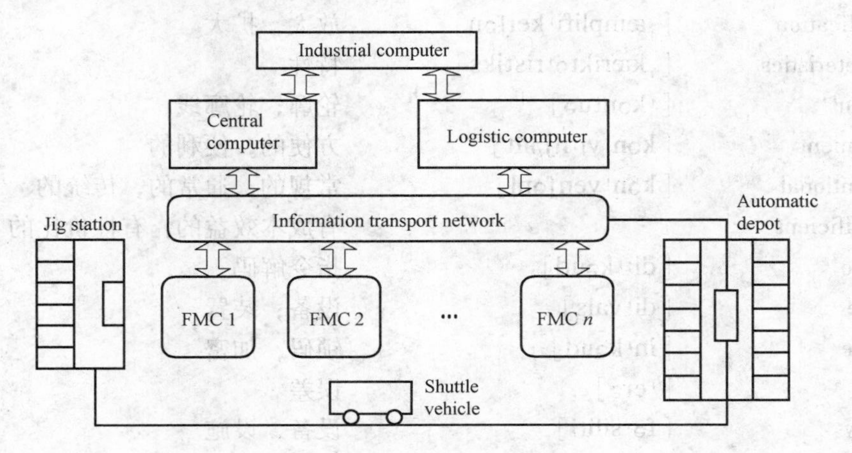

Figure 1-24 Construction of a FMS

(3) Computer integrated manufacturing system (CIMS) It can be regarded as a high efficiency FMS. Computer controls production procedure through ordering, design, machining, sales, etc. to realize synthetic automatic manufacturing rather than being restricted in machining (see Figure 1-25).

计算机集成制造系统（CIMS）是一种高效率柔性集成制造系统，应用最新的计算机技术控制，包括从订货、设计、工艺、制造、检验、销售到用户服务的全部过程，不仅实现了物料流自动化，还实现了信息流（管理）的自动化，如图 1-25 所示。

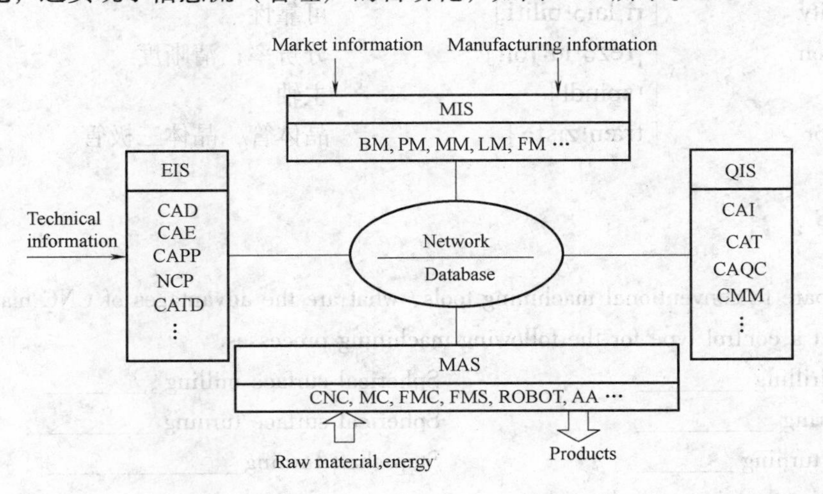

Figure 1-25 Construction of a CIMS

Glossary

accessory	[ækˈsesəri]	附件，配件，附属物
adaptability	[ˌədæptəˈbiliti]	适应性；灵活性
amplification	[ˌæmplifiˈkeiʃən]	放大，扩大
characteristics	[ˌkæriktəˈristiks]	特性
contour	[ˈkɔntuə]	轮廓；轮廓线
convenient	[kənˈviːnjənt]	方便的，便利的
conventional	[kənˈvenʃənl]	常规的，通常的，传统的
cost-efficient		有成本效益的；有性价比的
decode	[diːˈkəud]	指令解码
device	[diˈvais]	设备；装置
encode	[inˈkəud]	锁码；加密
error	[ˈerə]	误差
facility	[fəˈsiliti]	设备，设施
feedrate		进给率
flexibility	[ˌfleksəˈbiliti]	挠性，柔（韧）性；适应性，灵活性
loop	[luːp]	环路；环线
manual	[ˈmænjuəl]	手工的；用手操作的
mechanism	[ˈmekənizəm]	机械装置；机构
microprocessor	[ˌmaikrəuˈprəusesə]	（计算机）微处理器
performance	[pəˈfɔːməns]	（机器等的）性能
practicability	[ˌpræktikəˈbiliti]	实用性，可行性
reliability	[riˌlaiəˈbiliti]	可靠性
resolution	[ˌrezəˈluːʃən]	分辨率；清晰度
spindle	[ˈspindl]	主轴
transistor	[trænˈzistə]	晶体管，晶体三极管

Exercises

1. Compare to conventional machining tools, what are the advantages of CNC machine tools?

2. Select a control type for the following machining processes.

Holes drilling ________ Spherical surface milling ________

End boring ________ Spherical surface turning ________

Thread turning ________ Step shaft turning ________

(A) point-to-point control

(B) straight-line control

(C) 2D contour control

(D) 3D contour control

3. Draw the work principle diagrams of open loop control, closed loop control and semi-closed loop control. Explain why semi-closed loop control has the widest application today?

4. What is "pulse equivalent"? What is the unit of it?

5. If you are assigned to purchase machine tools for producing valve covers (15,000pieces/year), will you select CNC machine tools or conventional ones? Why?

6. Try to design machining process for the gearbox housing (shown in Figure 1-26) by using conventional machine tools and by using a milling center.

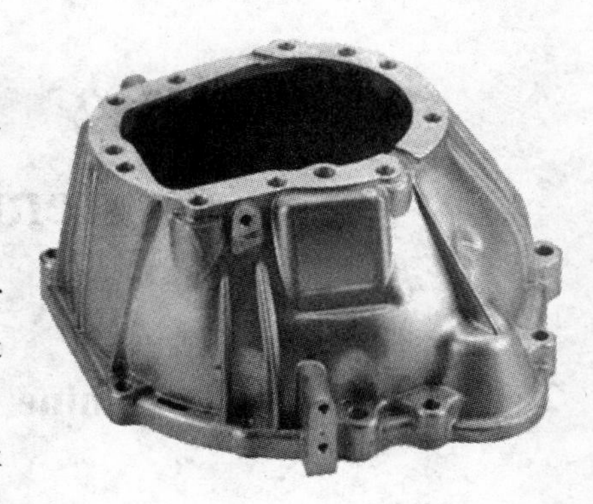

Firgure 1-26 Gearbox housing

Chapter 2 Motions and Coordinate System of Machine Tools

2.1 Motions of machine tools

2.1.1 Purpose of motions

The purpose of machine tool motions is to remove materials from workpieces to obtain required shapes, dimensions and surface quality. Tool motions, workpiece motions, and cutting motions are essential.

机床刀具运动可将工件或坯料上多余的材料层切除，从而使工件获得所需要的几何形状、尺寸和表面质量。刀具运动、工件运动和切削运动是机床的基本运动。

2.1.2 Surface forming motion

The surface forming motion is the relative cutting motion between the tool and the workpiece. Types and quantity of surface forming motion determine the shape of a workpiece, machining methods and tool construction. The surface forming motion consists of the primary motion and the feeding motion. Figure 2-1 shows a screw turning process, which is a typical surface forming motion.

表面成形运动是通过刀具与工件间的相对运动直接进行切削的过程，它决定工件的形状、采用的加工方法和刀具结构。表面成形运动由主运动与进给运动组成，如图 2-1 所示。

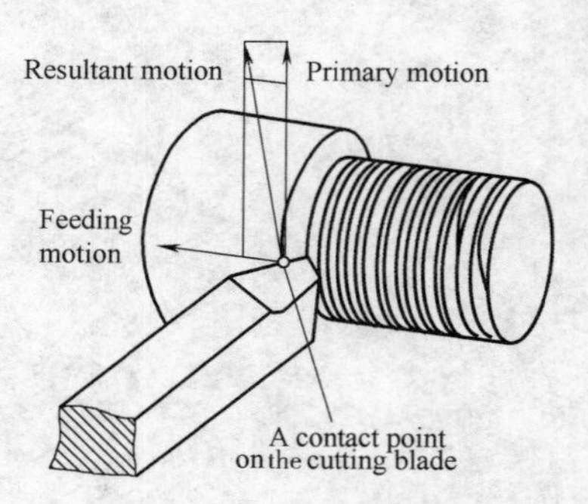

Figure 2-1 Screw turning

1. Primary motion

The primary motion causes relative motion between a cutting tool and a workpiece, and removes material from the workpiece.

Examples of primary motions:

1) The rotary motion of a workpiece (turning).

2) The rotary motion of a cutting tool (drilling, milling, boring).

3) The reciprocating motion of a cutting tool (broaching, planing).

Note:

Any machining process must have, and only has one primary motion.

机床的主运动使刀具从工件上直接切除金属。任何切削过程有且只有一个主运动。

2. Feeding motion

The feeding motion causes a secondary relative motion between the cutting tool and the workpiece, and enables the tool to continuously remove material from the workpiece to form a desired shape.

Examples of feeding motions:

1) The axial and radial motion of a cutting tool (turning).

2) The step motion of a cutting tool or a workpiece (planing).

Note:

① The feeding motion can be performed by either cutting tools or workpieces. The worktable performs feeding motion in milling (see Figure 2-2).

② Two or more feeding motions can be co-existing (gear hobbing).

③ Zero feeding motion is possible (broaching, see Figure 2-3).

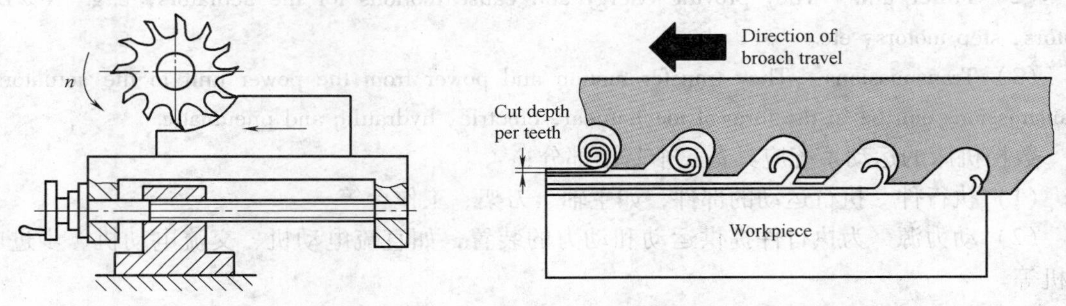

Figure 2-2 Milling   Figure 2-3 Broaching

机床进给运动使主运动能够持续不断地进行切削。在切削中可以有一个或多个进给运动（如滚齿），也可以不存在进给运动（如拉削）。

2.1.3 Assistant motion

All motions other than the surface forming motion are regarded as the assistant motion, including the plunge motion, the indexing motion, the alignment motion, fast moving, and other actions.

(1) The plunge motion Plunge a cutting tool into a workpiece to obtain a certain dimension.

(2) The indexing motion Periodical moving/rotating of a worktable or a tool post for machining a certain workpiece with symmetrical surfaces, e. g. multi-thread screws turning, polygon machining, etc.

(3) The alignment motion To obtain a correct relative position between a cutting tool and a workpiece, e. g. aligning drill to the hole center by moving the rocker arm of the driller.

(4) Fast moving and other actions Fast moving improves machining efficiency. Other actions include direction change, clamping, releasing, etc.

机床的辅助运动是指机床上除表面成形运动以外的所有运动，主要包括：

(1) 切入运动 为了获得一定的加工尺寸，使刀具切入工件的运动。

（2）分度运动　通过多工位工作台、刀架等实现周期性移位和转换，对工件进行顺序加工的运动。

（3）调整运动　使刀具和工件处于正确的相对位置的运动。

（4）快速移动及其他运动　快速移动为了提高生产率而快进和快退的运动。其他运动包括转向、装夹等。

2.2　Machine tool drive

To realize those necessary motions for machining, any CNC machine tool should have three basic units:

(1) Actuators　Their motions are the purpose of transmissions, e. g. spindles, tool posts, worktables, etc.

(2) Power unit　They provide energy and cause motions for the actuators, e. g. AC/DC motors, step motors, etc.

(3) Transmissions　They transfer motion and power from the power unit to the actuators. Transmissions can be in the form of mechanical, electric, hydraulic and pneumatic.

数控机床的运动系统应具备三个基本部分：

（1）执行件　执行运动的部件，如主轴、刀架、工作台等。

（2）动力源　为执行件提供运动和动力的装置，如直流电动机、交流电动机、步进电动机等。

（3）传动装置　连接动力源和执行件并传递运动和动力的装置，有机械、电气、液压、气压等几种类型。

2.2.1　Transmission connection

To obtain desired machine tool motions, a series of transmission components must be used to connect the power unit and the actuators, or to connect relevant actuators. The connection is known as the transmission connection.

为了得到所需的机床运动，需要通过一系列的传动件连接执行件和动力源，或者在相关的执行件之间建立连接，这种连接称为传动连接。

2.2.2　Transmission chain

A series of transmission components form a transmission connection, which is known as a transmission chain. Transmission chains are usually in two forms:

① Constant ratio and constant direction transmissions, e. g. gears, worm gears, ball screw-nut, etc.

传动比和传动方向不变的传动机构，如定比齿轮副、蜗杆副、滚珠丝杠副等。

② Variable ratio and variable direction transmissions, e. g. adjusting gear, clash gear, clutch gear shift, etc

传动比和传动方向可根据要求变化的传动机构，如交换齿轮变速机构、滑移齿轮变速机

构、离合器变速机构等。

Transmission chains can be in external and internal relations.

(1) External relation It connects the power unit and the actuators to drive the actuator running at a designed speed. Transmission ratio of external relation determines machining efficiency and surface roughness rather than shape of the workpiece.

外传动链连接动力源和执行件，使执行件以预定速度运动。其传动比只影响生产率或表面粗糙度，不影响成形运动。

(2) Internal relation It connects and regulates motions between the actuators. A precise transmission ratio is vital. Otherwise, the motion contour may not accurate. For internal relations, frictional transmissions (e. g. clutch) and non-constant ratio transmissions (e. g. chain drive) are unacceptable.

For example, screw turning (see Figure 2-4) requires a stringent transmission ratio between spindle rotation and tool post moving to ensure precise thread pitch.

内传动链是连接执行件与执行件之间的传动链，对执行件之间的相对速度或相对位移量有严格的要求，否则无法保证切削所需的正确运动轨迹。

在卧式车床上用螺纹车刀（图 2-4）车削螺纹时，连接主轴和刀架之间的螺纹传动链是一条对传动比有严格要求的内传动链，它能保证被加工螺纹的精确螺距。

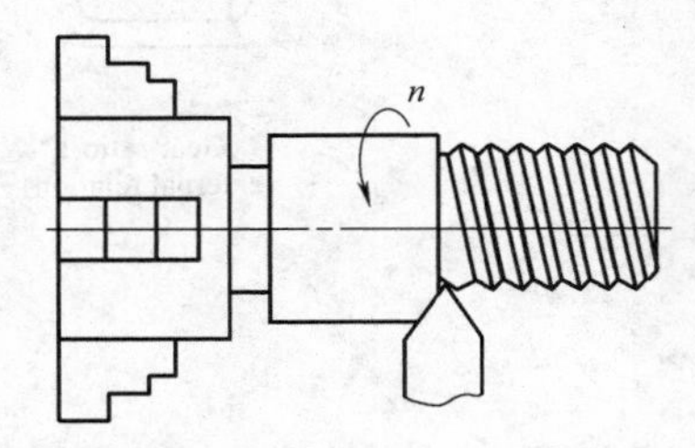

Figure 2-4 Screw turning

2. 2. 3 Transmission diagram

The transmission diagram briefly represents relative motion paths of surface forming by a group of symbols (see Figure 2-5).

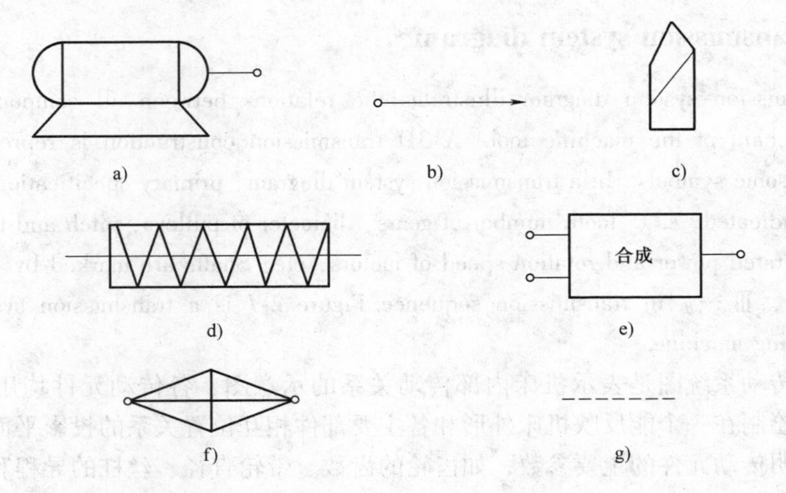

Figure 2-5 Symbols used in transmission diagram

a) Motor b) Spindle c) Turning tool d) Hob e) Combined unit f) Variable ratio g) Constant ratio

传动原理图通过一些简明的标准符号来表示与表面成形有关的运动和传动路径，如图 2-5 所示。

In Figure 2-6, component u_v determines the transmission ratio between the motor and the spindle, and component u_f determines the transmission ratio between the spindle and the lead screw to obtain required thread pitch.

For CNC machine tools, moving components may be driven by their own motors. The internal relations are coordinated by CNC device rather than mechanical connections.

数控机床各运动部件通常由各自的电动机驱动，内联系传动链由数控系统进行协调与控制。

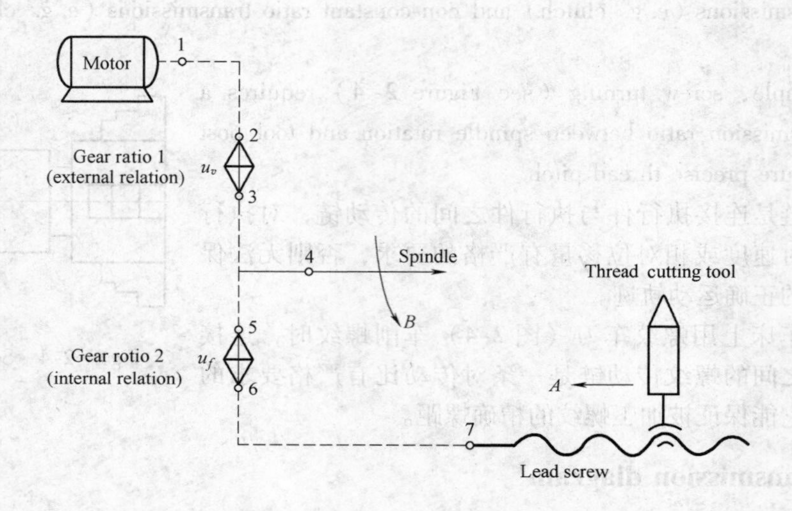

Figure 2-6 Transmission diagram of screw turning

2.2.4 Transmission system diagram

A transmission system diagram illustrates the relations between all components in a main transmission chain of the machine tool. A 3D transmission construction is represented by a 2D diagram and some symbols. In a transmission system diagram, primary specifications of components are usually indicated, e. g. tooth number of gears, diameter of pulleys, pitch and thread number of lead screws, rated power and rotation speed of motors, etc. Shafts are marked by Roman numerals (i. e. Ⅰ, Ⅱ, Ⅲ…) in transmission sequence. Figure 2-7 is a transmission system diagram of XK5750. milling machine.

机床的传动系统图是表示机床内部传动关系的示意图，各传动元件均用简单的符号表示，并展开绘制在一个能反映机床外形和各主要部件相互位置关系的投影平面内。传动系统图中通常注明传动元件的主要参数，如齿轮的齿数、带轮直径、丝杠的导程和线数、电动机的转速和功率等。传动轴从原动机开始，按传动顺序依次用罗马数字（Ⅰ、Ⅱ、Ⅲ…）编号。图 2-7 所示为 XK5750 铣床的传动示意图。

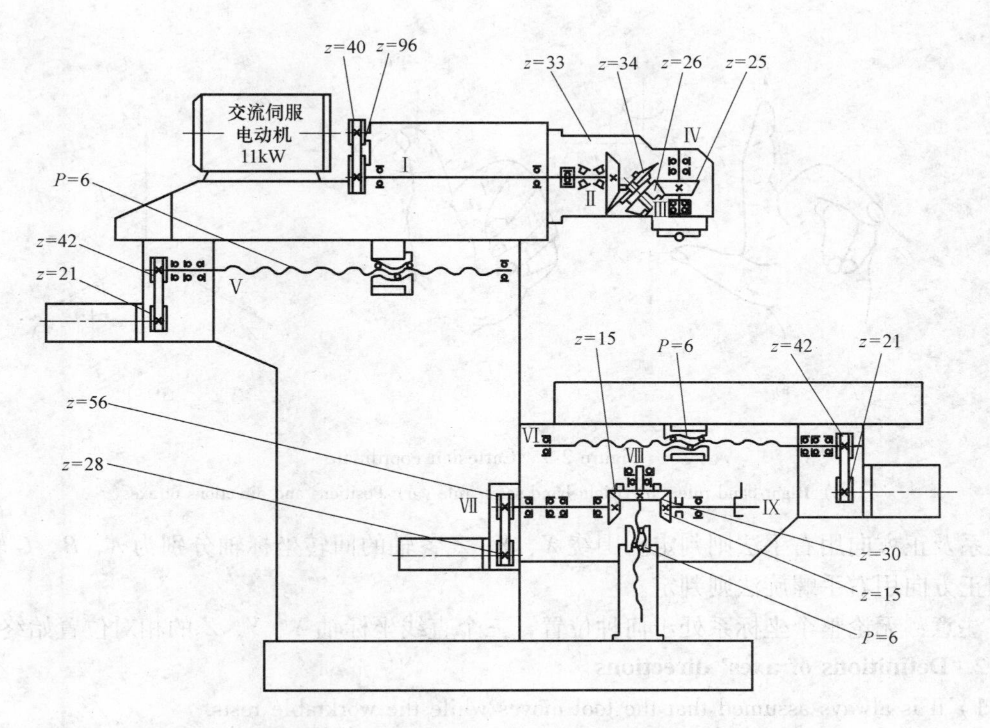

Figure 2-7 Transmission system diagram of XK5750 milling machine

2.3 Coordinate system for CNC machine tools

2.3.1 Standard coordinate system and direction

1. ISO coordinate system

Cartesian coordinate system has been defined as a standard for CNC machine tools by International Organization for Standardization (ISO). The adjective Cartesian refers to the French mathematician and philosopher René Descartes whose Latinized name was Cartesius (see Figure 2-8).

In this system, linear axes are defined as X, Y, and Z. The relative positions and directions of the axes are determined by right-hand rule (see Figure 2-9 a). The polar axes A, B, and C are respectively rotating about linear axes X, Y, and Z. The directions of polar axes are determined by right-hand screw rule (see Figure 2-9 b).

Note: The relative position of axes X, Y, and Z is always the same. Otherwise, it is a nonstandard coordinate system.

Figure 2-8 René Descartes (1596 - 1690)

国际标准化组织统一规定标准坐标系采用右手直角笛卡儿坐标系。其中，X、Y、Z 分别表示三个直线坐标轴，三者

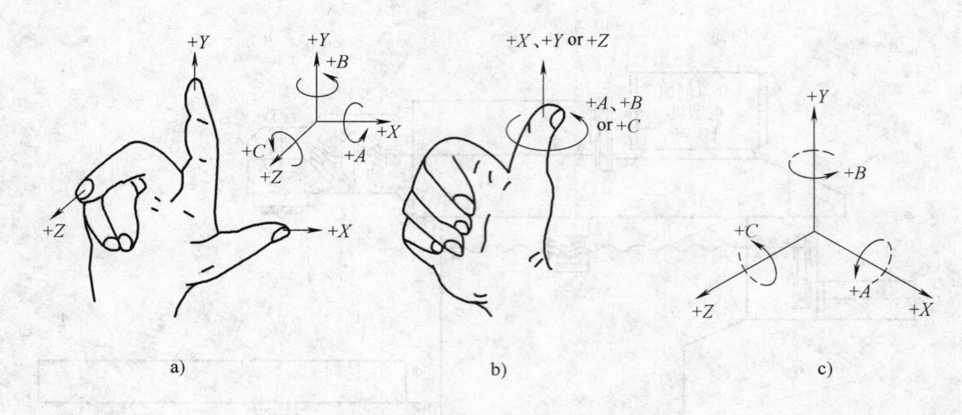

Figure 2-9 Cartesian coordinate

a）Right-hand rule b）Right-hand screw rule c）Positions and directions of axes

的关系及正方向用右手法则判定。围绕 *X*、*Y*、*Z* 各轴的回转坐标轴分别为 *A*、*B*、*C* 轴，它们的正方向用右手螺旋法则判定。

注意：无论整个坐标系处于何种位置，三个直线坐标轴 *X*、*Y*、*Z* 的相对位置始终不变。

2. Definitions of axes' directions

1）It is always assumed that the tool moves while the worktable rests.

2）The direction that makes the tool and the workpiece apart is defined as the positive direction.

3）The rotation that makes a right-hand thread screws into the workpiece is defined as the positive rotation.

坐标轴方向的判定原则：

1）对于工艺设计或编程，始终假定刀具运动而工件静止。

2）刀具远离工件的运动方向为坐标轴的正方向。

3）机床主轴按照右旋螺纹旋入工件的方向是旋转运动的正方向。

3. Axes of machine tools

（1）*Z*-axis *Z*-axis is parallel to the axis of the spindle. For those machine tools that have no spindle（e. g. planers）, *Z*-axis is vertical to the clamping plane of the worktable. When the tool moves away from the workpiece along *Z*-axis, the direction is defined as the positive direction of *Z*-axis.

Z 轴平行于主轴轴线，对于没有主轴的机床（如刨床），则规定垂直于工作台工件装夹面的轴为 *Z* 轴。*Z* 轴的正方向是使刀具远离工件的方向。

（2）*X*-axis *X*-axis is usually parallel to clamping plane.

The direction of *X*-axis depends on machine tool types. For horizontal machine tools（Figure 2-10a）, positive *X*-axis is pointing rightwards when looking from the spindle to the workpiece. For vertical machine tools（Figure 2-10b）, positive *X*-axis is pointing leftwards when looking from the spindle to the column. For machine tools that have no rotary cutting motion, positive *X*-axis is parallel to the direction of main cutting force.

X 轴平行于工件的装夹平面。对于卧式机床，从主轴向工件看，*X* 轴的正方向指向右边（图 2-10a）；对于立式机床，从主轴向立柱看时，*X* 轴的正方向指向左边（图 2-10b）。对于没有旋转刀具或旋转工件的机床，*X* 轴平行于主要的切削力方向，且以该方向为正方向。

（3）*Y*-axis *Y*-axis is vertical to both *Z* and *X*-axis. The positive direction of *Y*-axis is defined by right hand rule with known *Z* and *X*-axis.

Y 轴垂直于 *Z* 轴和 *X* 轴，可根据已知的 *X*、*Z* 轴正方向用右手法则确定 *Y* 轴的正方向。

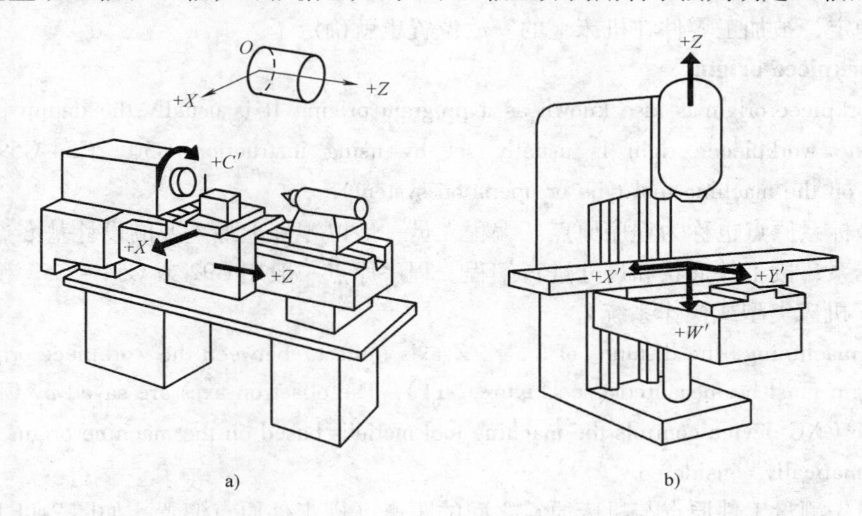

Figure 2-10 Axes of machine tools

a）Axes of a horizontal lathe b）Axes of a vertical milling machine

2.3.2 Coordinate systems of machine tools

（1）Machine origin The machine origin is also known as absolute origin or home position. Different machines have different machine origin, for milling, it is a physical position on the machine tool set during manufacturing. Usually, it locates at positive limits of *X*, *Y*, and *Z*-axis. All measurements of the machine tool are based on the machine origin.

机床原点也称为绝对原点，是制造商设置在机床上的一个物理位置，机床不同，其机床原点也不相同，例如铣床的机床原点一般位于三个坐标轴的正极限位置。它是测量机床运动坐标的起始点。

（2）Reference point The reference point is determined by stroke limiters. The reference point and the machine origin of a machine tool have a fixed relative position that is precisely measured by manufacturer, and usually are not on the same position.

For most machining centers, the spindle must return to the reference point before tool changing.

机床参考点是机床上用行程开关设置的一个物理位置，出厂之前由制造商精密测量确定。它与机床原点的相对位置是固定的，且一般不在同一点。

对于加工中心，机床参考点通常为自动换刀点。

2.3.3 Workpiece coordinate system

1. Workpiece coordinate system

The workpiece coordinate systemis also known as a program coordinate. It gives a reference position of each point of a workpiece. Once the workpiece coordinate system is established, position of the workpiece on the worktable is determined as well.

工件坐标系也称为编程坐标系，它使得零件上的所有几何元素都有确定的位置。工件坐标系一旦确定，被加工零件在机床上的安放位置也就确定了。

2. Workpiece origin

The workpiece origin is also known as a program origin. It is usually the datum point of the workpiece. the workpiece origin is usually set by using instruction G50, G54-G59, or G92 (depending on the machine tool type or operation system).

工件坐标系原点也称为程序原点。编程人员一般以零件上最重要的设计基准点为原点建立工件坐标系，按工件形状和尺寸进行编程。程序原点一般用 G92 或 G54 ~ G59 和 G50 来设置（取决于机床类型或操作系统）。

Before machining, the distance of *X*, *Y*, *Z*-axis (offset) between the workpiece origin and the machine origin must be measured (see Figure 2-11). The offset on axes are saved by CNC device. Although the CNC device controls the machine tool motions based on the machine origin, the offsets will be automatically considered.

加工前先测量工件原点与机床原点之间的距离（即工件原点偏置，如图 2-11 所示），并将该偏置值存入数控系统。加工时，数控系统会自动将偏置值加到工件坐标系上对工件进行加工。

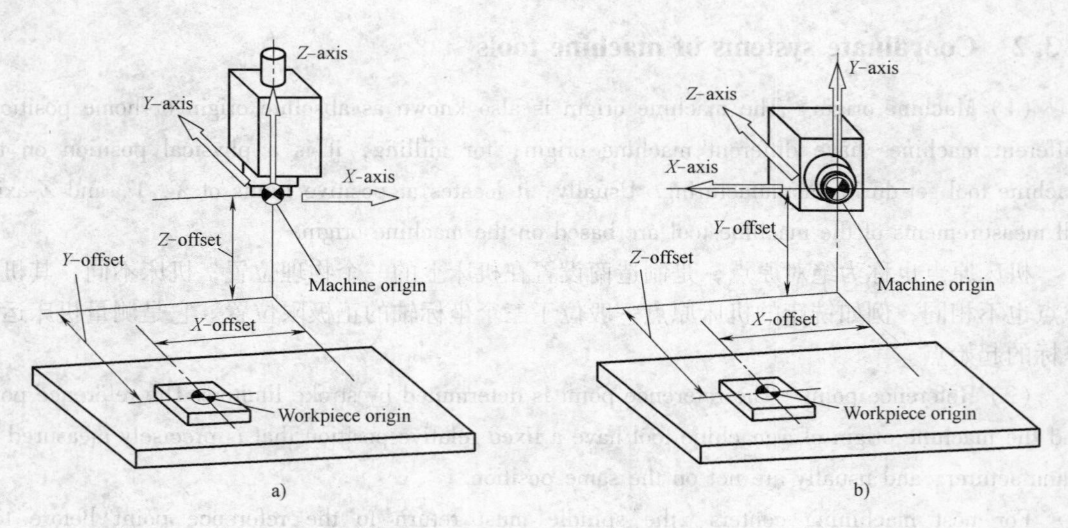

Figure 2-11 Coordinate systems of machine tools

a) Coordinate system of the vertical machine tool b) Coordinate system of the horizontal machine tool

Glossary

alignment	[ə'lainmənt]	调准，校直
assistant	[ə'sistənt]	起辅助作用的事物
boring	[ˈbɔːriŋ]	钻孔；镗孔
broaching	[brəutʃiŋ]	拉削（加工）
Cartesian coordinate		笛卡儿坐标
coordinate	[kəuˈɔːdneit]	坐标的
drilling	[ˈdriliŋ]	钻孔
indexing	[ˈindeksiŋ]	分度
milling	[ˈmiliŋ]	铣削（加工）
origin	[ˈɔːridʒin]	（坐标）原点
planing	[plæniŋ]	刨削（加工）
right hand rule		右手法则
thread	[θred]	螺纹
turning	[ˈtəːniŋ]	车削（加工）

Exercises

1. Indicate the primary motion and the feeding motion for a 2D contouring milling.
2. Could a friction clutch be used in the transmission chain of thread turning? Why?
3. Determine if the following coordinate systems (shown in Figure 2-12) are standard ones.

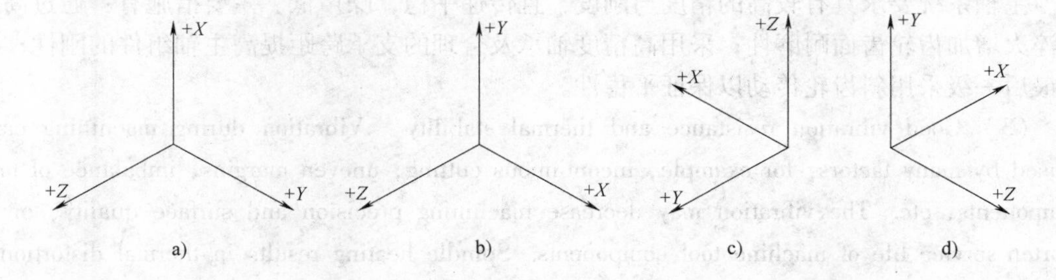

Figure 2-12 Coordinate systems

Chapter 3 Typical Components of CNC Machine Tools

3. 1 Spindle system of CNC machine tools

3. 1. 1 The requirements of the spindle system

As one of the forming motions of a machine tool, spindle motion precision determines machining accuracy. To provide high efficiency of CNC machine tools, the spindle system is expected to have:

(1) Wider rotation speed range and continuous transmission For ensuring the optimum cutting rate to improve efficiency, accuracy and surface quality of machining, the spindle of the CNC machine tool usually has a reduction rate of 1:(100-1000) for a constant torque, and 1:10 for a constant power.

数控机床能在较大的转速范围内实现无级调速，一般要求主轴具备 1:(100 ~ 1000) 的恒转矩调速范围和 1:10 的恒功率调速范围。

(2) High precision and rigidity, smooth transmission and silent All the gears in spindle transmission chain have been treated by high frequency quenching to improve anti-abrasion characteristic. High quality bearings and optimum support span improve precision and rigidity of the spindle system. Helical gears are used at the end of the transmission chain for smoothness.

主轴系统要求具有较高的精度与刚度，且传递平稳，噪声低。主要措施有：通过高频感应淬火增加齿轮齿面耐磨性；采用高精度轴承及合理的支承跨距提高主轴组件的刚性；传动链最后一级采用斜齿轮传动以保证平稳性。

(3) Good vibration resistance and thermal stability Vibration during machining can be caused by many factors, for example, incontinuous cutting, uneven margins, imbalance of motion components, etc. The vibration may decrease machining precision and surface quality, or even shorten service life of machine tool components. Spindle heating results in thermal distortion that reduces transmission efficiency and increases machining errors. Therefore, spindle components must have high inherent frequency. Reasonable fitting clearance and circulating lubrication are also necessary.

数控机床加工时，可能由于断续切削、加工余量不均匀、运动部件不平衡以及切削过程中的自振等原因引起的冲击力的干扰会使主轴产生振动，从而影响加工精度和表面粗糙度，严重时甚至可能破坏刀具和主轴系统中的零件，使其无法工作。主轴系统发热使其中的零部件产生热变形，降低传动效率，破坏零部件之间的相对位置精度和运动精度，造成加工误差。为此，主轴组件要有较高的固有频率，实现动平衡，保持合适的配合间隙并进行循环润滑等，以获得良好的抗振性和热稳定性。

(4) *C*-axis control is a must for turning center.

在车削中心上，要求主轴具有 *C* 轴控制功能。

(5) Precise orientation of the machining center The spindle is required to precisely stop at a certain position for automatic tool changing. The ability of accurate stop is known as spindle orientation.

在加工中心上，要求主轴具有高精度的准停功能。在加工中心上自动换刀时，主轴必须停止在一个固定不变的方位上，以保证换刀位置的准确以及某些加工工艺的需要，即要求主轴具有高精度的准停功能。

(6) Constant linear velocity control When machining by a lathe or a grinding machine, to obtain an even surface quality, a constant relative linear velocity between the cutting tool and the workpiece is required. For example, when the cutting tool is approaching the center line of the workpiece, the spindle rotation speed would correspondingly increase to maintain the same contact speed between the tool nose and cutting surface.

利用车床或磨床进行回转类工件加工时，为了保证直径不同的各处表面粗糙度一致，随着切削直径的变化，需要不断调整主轴转速，保持刀具切削的线速度为恒定值。

3.1.2 Transmission mode of CNC machine tools

To provide a wide rotation speed range, spindle transmission configurations are generally in four types (see Figure 3-1):

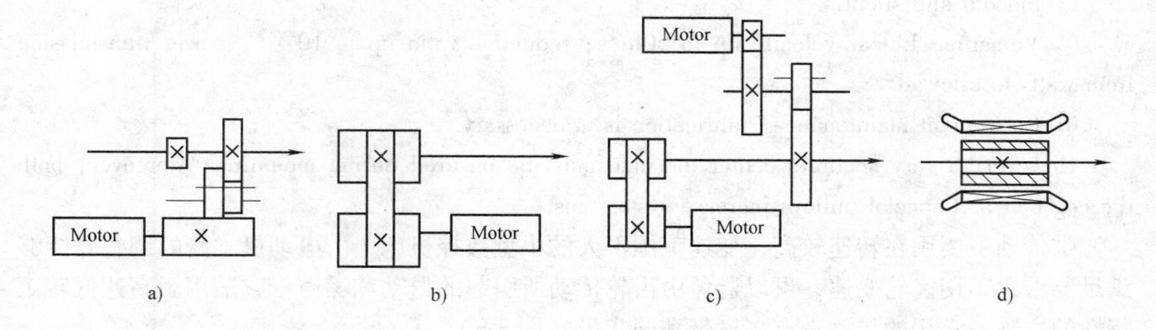

Figure 3-1 Spindle transmissions of CNC machine tools

a) Gear shifting b) Belt drive c) Direct drive d) Variable speed motor direct drive

1. Gear shifting (see Figure 3-1a)

It is usually applied on large size machine tools for adequate torque at low rotation speed. Shifting motion is executed by hydraulic forks or hydraulic cylinders.

A typical three-position hydraulic gear shifting mechanism is shown in Figure 3-2.

Status①: Hydraulic oil injects into cylinder 1 to push the gear into left position.

Status②: Hydraulic oil injects into cylinder 5 to push the gear into right position.

Status③: Hydraulic oil injects into cylinder 1 & 5 with the same pressure. The cross sectional area at the left end of the piston bar is smaller than that at the right end. However, with the bush, the total cross sectional area is larger than that of the right end, therefore, the gear is kept at middle position.

带有变速齿轮的主轴传动是大中型数控机床较常采用的配置方式，通过少数几对齿轮传动，扩大变速范围，确保低速时有较大的转矩，以满足主轴输出转矩特性的要求。滑移齿轮的移位大多采用液压拨叉或直接由液压缸驱动齿轮来实现。图 3-2 所示是三位齿轮液压拨叉的工作原理图。

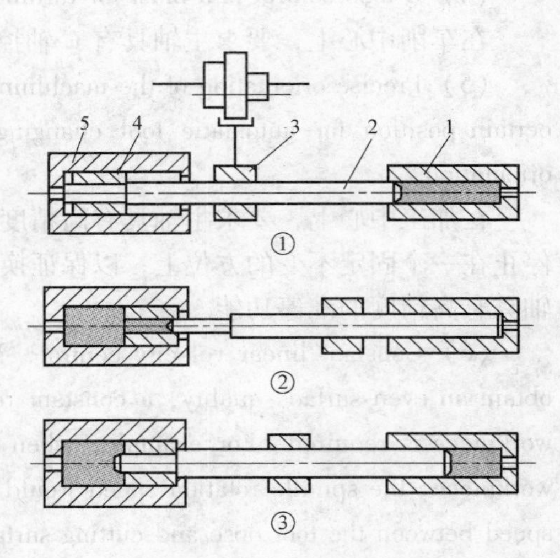

Figure 3-2　Spindle transmissions of CNC machine tools

1、5—Cylinder　2—Piston bar　3—Fork　4—Bush

2. Belt drive (see Figure 3-1b)

Belt drive is not designed for large torque transmission. It is applied on small size machine tools with high speed but narrow speed range. companed to gear drive, belf drive produces less vibration and noise. Cog belts (see Figure 3-3) are widely used in CNC machine tools.

Compared to flat belts and V-belts, cog belts have these advantages:

① Transmission efficiency reaches 98%.

② No slipping thus accurate reduction ratio.

③ Smooth and silent.

④ Versatile. Linear velocity up to 50m/s; reduction ratio up to 10: 1; power transmission from watts to kilowatt.

⑤ Convenient maintenance; lubrication is unnecessary.

However, a very accurate center distance must be ensured during mounting. Moreover, both the cog belt and special pulleys increase system cost.

带传动主要用在转速较高、变速范围不大的小型数控机床上。电动机本身的调整就能够满足要求，不用齿轮变速，可以避免由齿轮传动所引起的振动和噪声。它适用于高速低转矩特性的主轴。常用的传动带有多楔带和同步带（图 3-3）。

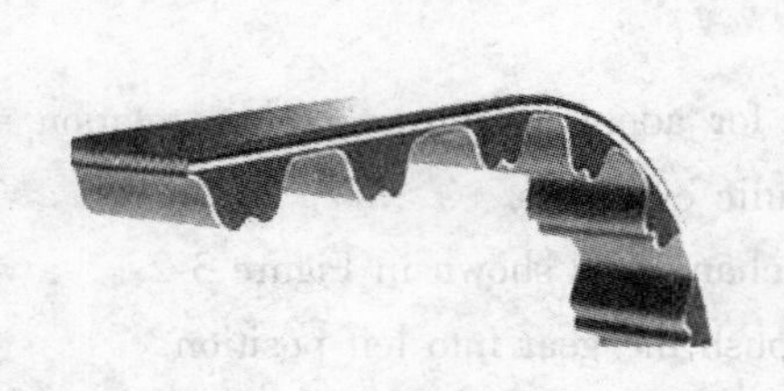

Figure 3-3　Cog belt and cog belt drive

多楔带的横向断面呈多个楔形，兼有 V 带和平带的优点，运转时振动小、发热少、运转平稳、重量小，可在 40m/s 的线速度下使用。

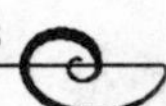

同步带传动是一种综合了带传动和链传动优点的新型传动方式。带的工作面及带轮外圆上均制成齿形，通过带轮与轮齿相嵌合，进行无滑动的啮合传动。与一般带传动相比，同步带传动具有如下优点：

① 传动效率高，可达98%以上。

② 无滑动，传动比准确。

③ 传动平稳，噪声小。

④ 使用范围较广，速度可达50m/s，速比可达10左右，传递功率由几瓦至数千瓦。

⑤ 维修保养方便，不需要润滑。

但是，同步带安装时中心距要求严格，而且带与带轮的制造工艺复杂，成本较高。

3. Double motor drive (see Figure 3-1c)

The spindle can be driven through belt drive with one motor at high rotation speed but low torque, or through gear drive with another motor at low rotation speed but high torque.

The configuration sufficiently utilizes motor power and expands constant power range. However, it is wasteful since only one motor is allowed to work at one time.

用两个电动机分别驱动主轴传动是一种混合传动方式。其中一个电动机通过带传动实现高速低转矩；另一个电动机通过齿轮进行降速，扩大变速范围，从而实现低速高转矩。这种配置可避免低速时转矩不够且电动机功率不能充分利用的问题。但是两个电动机不能同时工作。

4. Direct drive by variable speed motor (see Figure 3-1d)

In this configuration, the spindle is direct connected to the motor (see Figure 3-4). It has simplified structure and improved spindle rigidity, but its output torque is limited. Moreover, motor heating affects system accuracy.

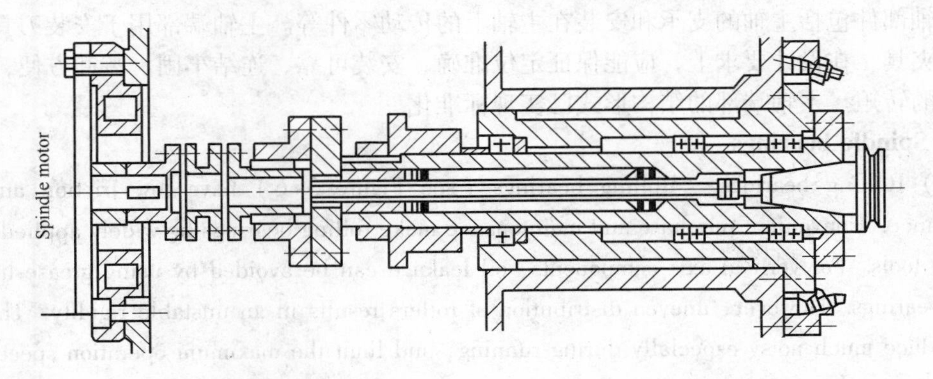

Figure 3-4 Rigid connection between cog belt and cog belt drive

An innovative spindle-integrated motor has been developed recently (see Figure 3-5). The compact structure reduces system inertia thus to improve start/stop respond (up to 180,000r/min), also reduce vibration and noise.

Since heat during working may result in thermal deformation, temperature control and cooling are key technologies of the system.

由调速电动机直接驱动主轴传动如图3-4所示。它大大简化了主轴箱体与主轴的结构，有效地提高了主轴部件的刚度，但主轴输出的转矩小，电动机发热对主轴的精度影响较大。

近年来出现了一种新式的内装电动机主轴，即主轴与电动机转子合为一体。其主轴组件结构紧凑、重量和惯量小，可提高起动、停止的响应特性，并利于控制振动和噪声；缺点是电动机运转产生的热量易使主轴产生热变形。因此，温度控制和冷却是使用内装电动机主轴的关键。

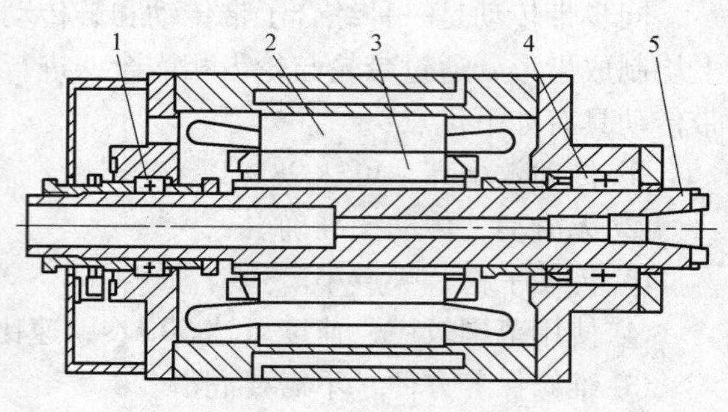

Figure 3-5 Spindle transmissions

1—rear bearing 2—Stator 3—Rotor
4— Front bearing 5—Spindle

3.1.3 Spindle components

Spindle components include spindle support and transmission parts.

1. End structure of the spindle

The spindle end carries a cutting tool or a clamp, therefore, some design requirements are expected:

1) Precise positioning.

2) Reliable clamping.

3) Firm connection.

4) Convenient fixing.

5) Be able to transmit enough torque.

Spindle end structures have been standardized.

主轴部件包括主轴的支承和安装在主轴上的传动零件等。主轴端部用于安装刀具或夹持工件的夹具，在设计要求上，应能保证定位准确、安装可靠、连结牢固、装卸方便，并能传递足够的转矩。主轴端部的结构形式已实现标准化。

2. Spindle bearings

(1) Rolling bearings Rolling bearings (see Figure 3-6) have low friction and stable performance. Convenient purchase and maintenance make rolling bearings be widely applied on CNC machine tools. For vertical axis components, oil leakage can be avoided by using grease-lubricated rolling bearings. However, uneven distribution of rollers results in an unstable rigidity. The rollers also produce much noisy especially during running, and limit the maximum operation speed.

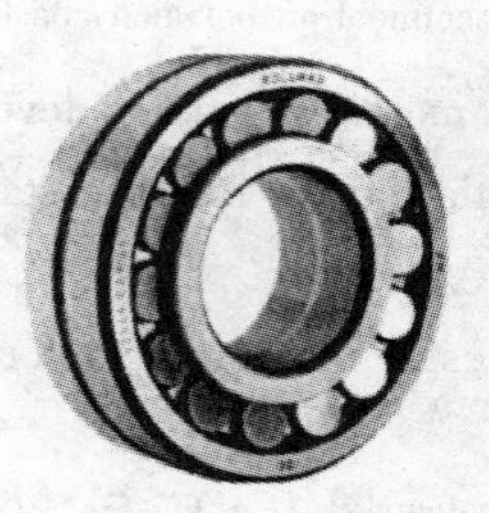

Figure 3-6 Rolling bearings

滚动轴承摩擦阻力小，可以预紧，润滑维护简单，选购维修方便，因此在数控机床上被广泛采用。但其噪声大，滚动体数目有限，刚度是变化的，抗振性略差，并且对最高转速有很大的限制。对于大多数立式主轴和装在套筒内能够作轴向移动的主轴，使用脂润滑滚动轴承可以避免漏油。

Mechanical characteristics of rolling bearings might be affected by mounting. Figure 3-7 illustrates three mounting types of a pair of angular contact ball bearings. Type a can withstand axial force in both direction and has the best stability. Type b has the worst stability. Type c can bear double axial force in one direction, and has the average stability.

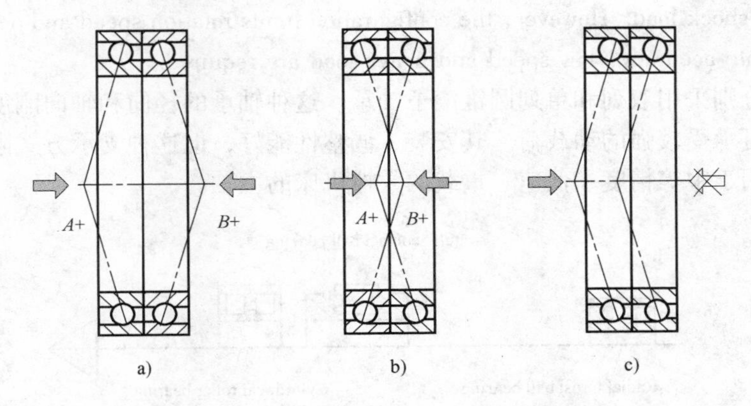

Figure 3-7 Mounting types of angular contact ball bearings

a) Indirect b) Direct c) Same directional

(2) Static pressure bearing It is commonly used in CNC machine tools to withstand pressure force by a thin oil film. The performance of a static pressure bearing is independent to the spindle rotation speed because the pressure is generated by a hydraulic system. Since no mechanical contact between rotating components and bearings, it can realize a friction-free operation. Static pressure bearing gives the machine tool ideal motion accuracy. However, the complex construction and an additional hydraulic system significantly increase the cost.

在数控机床上最常使用的滑动轴承是静压滑动轴承。静压滑动轴承的油膜压力是由液压缸从外界供给的，它与主轴是否转动、转速高低均无关（忽略旋转时的动压效应）。它的承载能力不随转速变化，运行时无磨损，起动和运转时摩擦阻力力矩相同。因此静压轴承的刚度大，回转精度高，但静压轴承需要一套液压装置，成本较高。

3. Spindle bearing configurations

Firgure3-8 shows the configurations of spindle bearings. The configurations are usually in three types, depending on the rotation speed and rigidity requirements.

① Angular contact ball bearing and cylindrical roller bearing for front support; Radial thrust ball bearing for rear support.

The configuration reinforces the spindle rigidity to withstand large cutting force. It is applied on most machine tools.

前轴承采用双列短圆柱滚子轴承角接触双列向心推力球轴承组合，后轴承采用成对向心

推力球轴承。此配置可提高主轴的综合刚度，满足强力切削的要求，广泛用于各类数控机床主轴。

② A group of radial thrust ball bearings allow the spindle to run at high rotation speed (up to 4000r/min). The configuration is applied on where high speed, light loading and high accuracy are required.

前轴承采用一组高精度双列向心推力球轴承，以获得良好的高速性，主轴最高转速可达4000r/min。该配置的承载能力小，适用于高速、轻载、高精密的数控机床主轴。

③ Tapered roller bearings for both front and rear supports allow the spindle to withstand heavy load, especially shock load. However, the configuration limits rotation speed and accuracy. Applied on where moderate accuracy, low speed and heavy load are required.

前后轴承分别采用双列和单列圆锥滚子轴承。这种轴承的径向和轴向刚度高，能承受重载荷，尤其是可承受较强的动载荷。其安装、调整性能好，但这种支承方式限制了主轴转速和精度，通常用于中等精度、低速、重载的数控机床的主轴。

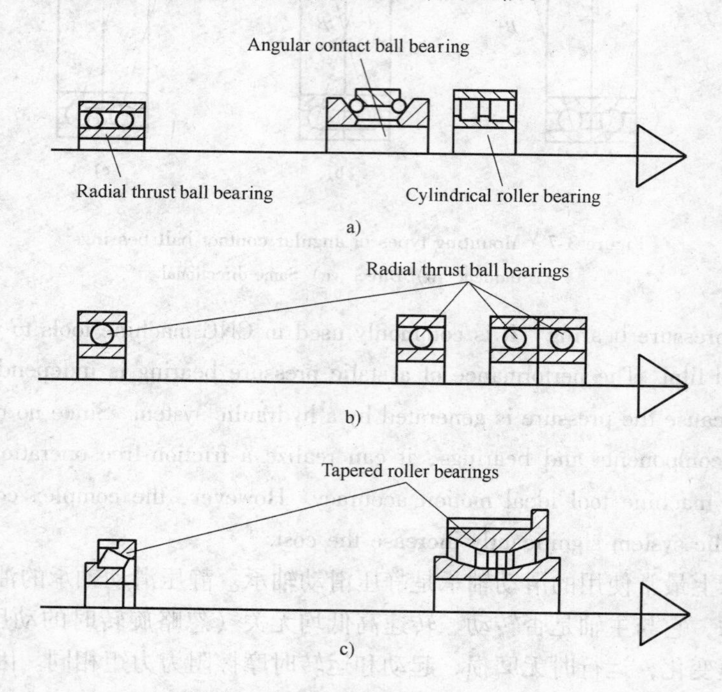

Figure 3-8 Configurations of spindle bearings

4. Clearance and preload

For rolling bearings, a clearance between rings and rollers results in load concentrating, i. e. only some of the rollers are withstanding load. It decreases rotational accuracy and loadability. Preload increases the number of rollers that effectively contact rings for the load. Preloading force varies with spindle working conditions between different machine tools. Therefore, bearing clearance adjusting mechanism is an important component in spindle construction.

It must be noticed that over-preloading will produce excessive friction between rings and rollers thus shortening service life of the bearings.

滚动轴承间隙较大时，载荷集中作用于受力方向上的少数滚动体上，导致轴承承载能力下降，旋转精度降低。将滚动轴承进行适当预紧，就可使承载的滚动体数量增多，受力趋向均匀。主轴轴承所需的预紧量取决于机床类型和主轴的工作条件。因此，主轴组件必须具备调整轴承间隙的机构。

需要注意的是预紧量不宜过大，否则会加剧轴承的摩擦，增加能耗，降低工作寿命。

5. Bearing accuracy

Four precision classes of bearings are used on spindle components:

Class E: Average precision.

Class D: Fine precision.

Class C: Super-precision.

Class B: Hyper-precision.

Usually, front bearing has higher precision than rear bearing though the same precision is acceptable. Class C or D is applied on front bearings for conventional precision machine tools, and Class D or E for rear bearings. Class B is applied on both front and rear bearings only for high precision machine tools.

主轴部件所用滚动轴承的精度有高级 E 级、精密级 D 级、特精级 C 级和超精级 B 级。前轴承的精度一般比后轴承的精度高一级，也可以用相同的精度等级。普通精度的机床通常前轴承用 C 或 D 级，后轴承用 D 或 E 级。特高精度的机床前后轴承均用 B 级。

3.1.4 Spindle orientation

Spindle orientation is also known as spindle positioning. For a milling/boring center, the torque of machining is delivered through key connection between the spindle and the tool handle. Therefore, when the spindle stops for tool changing, it must be in a certain angle to match the keyway on the tool handle (see Figure 3-9).

The other purpose of spindle orientation is to prevent the finished surface from scrape during the cutting tool withdrawing (see Figure 3-10).

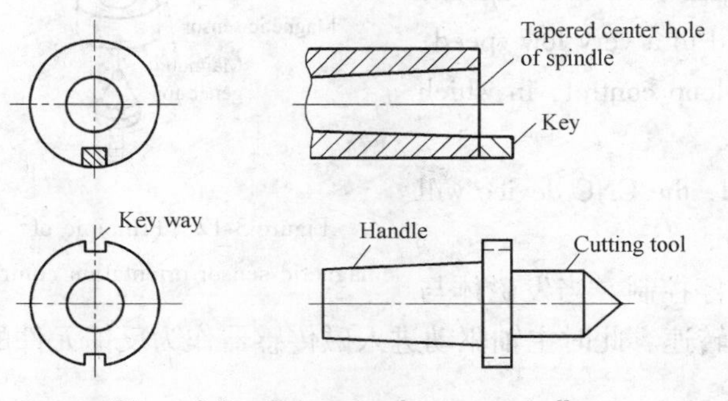

Figure 3-9 Connection between spindle and cutting tool

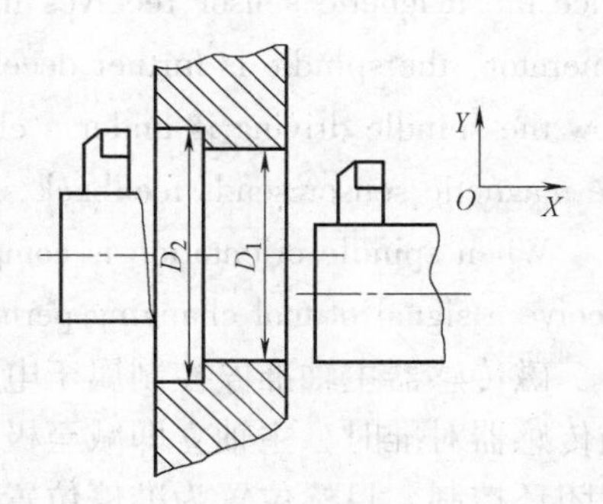

Figure 3-10 Cutting tool withdrawing from a stepped hole

主轴准停功能又称为主轴定位功能。数控镗床或铣床的切削转矩通常是通过主轴上的端面键和刀柄上的键槽来传递的，因此每次自动换刀时，都必须使刀柄上的键槽对准主轴的端面键（图 3-9），这就要求主轴具有准确定位于圆周上特定角度的功能。

当加工阶梯孔或精镗孔后退刀时，为防止刀具与小阶梯孔碰撞或拉毛已完成精加工的孔表面（图 3-10），必须先让刀，再退刀，因此刀具必须有定位功能。

1. Mechanical orientation control

Figure 3-11 shows the principle of a mechanical orientation control. When the control system receives an orientation signal, the spindle will be decelerated to a certain low speed. After a signal from the approacher is received by contactless switch, the spindle motor will be power off, and at the same time, spindle transmission will be cut-off (now the spindle is still rotating due to inertia). The cylinder pushes the pin against the positioning plate until the pin inserts into the V-gap. As a result, LS_2 is triggered, which allows tool changing action. Both LS_1 and LS_2 are logic switches. Spindle motor won't start unless LS_1 is triggered. The process of spindle orientation can be controlled by PLC.

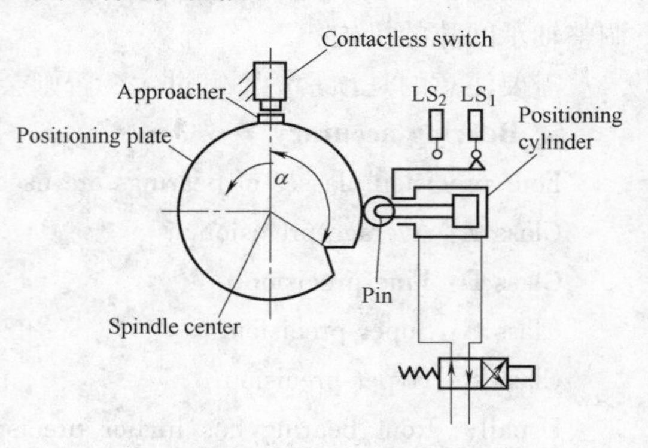

Figure 3-11 Construction of mechanical orientation control

图 3-11 所示为典型的 V 形槽轮定位盘机械准停原理示意图。接收到准停指令时，首先使主轴减至低速转动，当无触点开关有效信号被检测到后，立即使主轴电动机停转并断开主轴传动链，此时主轴电动机与主轴传动件因惯性继续空转，同时准停液压缸定位销伸出并压向定位盘。当定位盘 V 形槽与定位销正对时，由于液压缸的压力，定位销插入 V 形槽中。LS_2 为准停完成信号，LS_1 为准停解除信号。

2. Electric orientation control

(1) Magnetic sensor control (see Figure 3-12) Magnetic sensor control is driven by the spindle motor. When the orientation signal is received, spindle will be accelerated or decelerated to a certain speed (known as orientation speed). Once the magnetic sensor receives the signal from the magnetic generator, the spindle is further decelerated to a very low speed. Now the spindle driving is under a closed-loop control, in which the magnetic sensor sends feedback signals.

When spindle orientation is completed, the CNC device will receive a signal of tool changing permission.

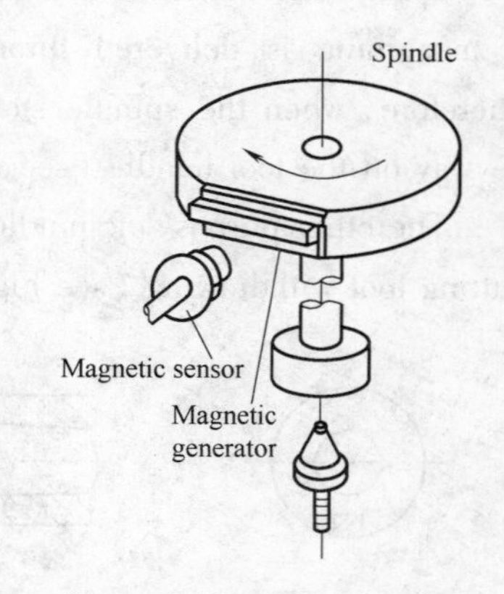

Figure 3-12 Principle of magnetic sensor orientation control

磁传感器主轴准停控制属于电气准停控制。当发磁体与磁传感器对准时，主轴立即减至极慢的转速。此时主轴驱动进入磁传感器作为反馈元件的位置闭环控制，目标位置为准停位置。

(2) Encoder orientation control Figure 3-13 illustrates the principle of encoder controlled spindle orientation. An encoder is fixed on the spindle or spindle motor for precise spindle angular position control. Compared to magnetic sensor control, angular position can be adjusted easily in an encoder controlled system. Work steps of the system are similar to that of magnetic sensor

orientation control.

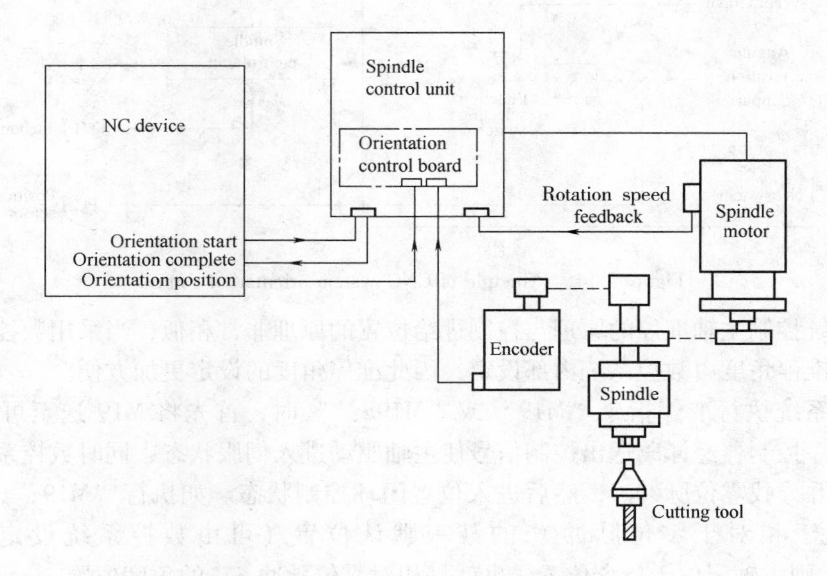

Figure 3-13 Principle of encoder orientation control

图 3-13 所示为编码器主轴准停控制原理图。可以采用主轴电动机内部安装的编码器信号（来自于主轴驱动装置），也可以在主轴上直接安装另外一个编码器。主轴驱动装置可以使主轴驱动处于速度控制或位置控制状态。准停角度由外部开关量设定。工作步骤与磁传感器主轴准停类似。

（3） CNC system orientation control The principle of CNC system orientation control is similar to that of feeding control. When orientaton is required, CNC system would shift the spindle from rotation speed control to angular position control（see Figure 3-14）. The stop position can be simply input through a program. Code M19 is defined for spindle orientation.

When instruction M19 is executed, PLC switches the spindle control unit to servo drive mode, and the spindle runs at a low speed under closed-loop control. Instruction "M19S**" is to set the reference zero. If only "M19" is executed, the reference zero is a default. For example:

M03	S1000	// Spindle runs at 1000 r/min
M19	S100	// Spindle stops at an angular position of 100°
S1000		// Spindle runs at 1000 r/min again
M19		// Spindle stops at a default angular position（100°）
M19	S200	// Spindle stops at an angular position of 200°

Notice: CNC system controlled by spindle orientation can only be applied on where the spindle is under closed-loop control, and the spindle must be able to work at servo-mode. Moreover, orientation precision may be affected by transmission error if the spindle is under semi-closed loop control.

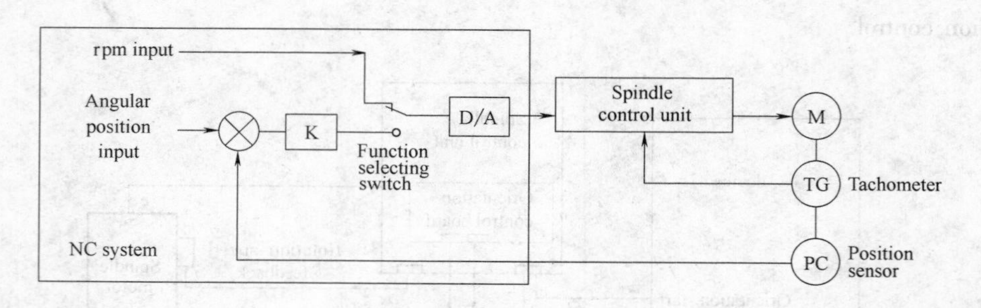

Figure 3-14　Principle of CNC system orientation control

数控系统控制主轴准停的原理与控制进给位置的原理非常相似，当采用数控系统控制主轴准停时，准停角度由数控系统内部设定，因此准停角度的设定更加方便。

当数控系统执行准停指令“M19”或“M19S**”时，首先将M19送至可编程序控制器，可编程序控制器经译码送出控制信号使主轴驱动进入伺服状态，同时数控系统控制主轴电动机降速并寻找零位脉冲C，然后进入位置闭环控制状态。如执行“M19;”，无S指令，则主轴定位于相对于零位脉冲C的某一默认位置（可由数控系统设定）。如执行“M19S**”，则主轴定位于指令位置，也就是相对零位脉冲S**的角度位置。

采用这种控制方式时需注意：数控系统必须具有主轴闭环控制功能，且主轴可在伺服状态下运转。此外，如果采用的是主轴半闭环控制，则传动误差可能对定位精度产生影响。

Most large CNC systems adopt electric orientation control modes, because they have following advantages:

① Simplified construction. Only a group of sensors on relevant rotating and static components are required.

② Lessened operation time. Electric orientation enables fast positioning even at high rotation speed. It finally lessens tool changing time.

③ Increased reliability. Service life of the control unit is extended due to elimination of complex mechanisms, switches and cylinders, and elimination of mechanical shock.

④ Reasonable cost-effectiveness.

3.1.5　Lubrication and sealing of spindle

1. Lubrication

For CNC machine tools, grease, oil circulation, oil mist and oil jet lubrication are usually applied.

(1) Grease lubrication　It is most commonly used for front bearings where it is difficult to arrange oil cooling system. The amount of grease should be no more than 10% of bearing space to prevent overheating. Effective sealing must be designed to keep cutting fluid or other lubricating oil out of the bearing.

油脂润滑是目前在数控机床的主轴轴承上，特别是在前轴承上最常用的润滑方式。如果主轴箱中没有冷却润滑油系统，那么后轴承和其他轴承一般也采用油脂润滑方式。

主轴轴承油脂封入量通常为轴承空间容量的10%，若油脂过多，会加剧主轴发热。油

脂润滑必须采用有效的密封措施，以防止切削液或其他润滑油进入轴承中。

(2) Oil circulating lubrication (constant temperature) Oil circulating lubrication is often applied on the rear bearings of the spindle. The circulation path of oil is shown in Figure 3-15. Oil temperature in reservoir is automatically maintained at a constant by a cooling system. Through filter, oil is pumped into lubricant distributor, and is divided into two paths, one of which flows through a helical groove over the surface of bearing bush to take heat away. The other flows into a sprayer, which injects the oil onto gears and bearings both for lubricating and cooling.

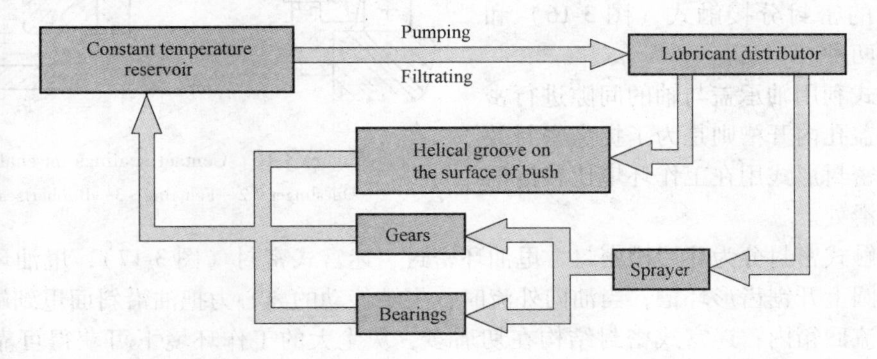

Figure 3-15 Constant temperature lubricant circulation

在数控机床主轴上，后轴承上采用油液循环的润滑方式比较常见。恒温油液循环润滑冷却方式如图 3-15 所示。由油温自动控制箱控制的恒温油液，经油泵到润滑分配箱，其中一路沿主轴前轴承套外圈上的螺旋槽流动，以带走主轴轴承所发出的热量；另一路通过主轴箱内的喷射装置把润滑油喷射到传动齿轮和传动轴轴承上。这种方式的润滑和降温效果都很好。

(3) Oil mist This method is to atomize lubricant by compressed air and spray onto mechanical components. Oil mist is suitable for high speed bearings because of its good heat absorption and no fluid whipping.

Notice: Oil mist is likely to escape thus result in pollution, therefore, it has been prohibited in many European countries.

油雾润滑方式是将油液经高压气体雾化后，从喷嘴喷到需润滑的部位的润滑方式。由于是雾状油液，其吸热性好，又无油液搅拌作用，因此常用于高速主轴轴承的润滑。但是油雾容易吹出，污染环境，目前欧洲有些国家已经禁止使用这种润滑方式。

(4) Oil jet lubrication Oil jet lubrication is especially designed for ultra-high speed spindle bearings. It only uses0. 2ml/hour of oil to prevent bearings from heating. Oil level switch in the reservoir and pressure switch in pipes will automatically cut off the spindle motor if oil level or pressure drops.

油气润滑方式是针对高速主轴而开发的新型润滑方式。它用极微量油（约 0. 2mL/h）润滑轴承，以抑制轴承发热。油箱中的油位开关和管路中的压力开关确保在油箱中无油或压力不足时能自动切断主电动机电源。

2. Sealing of the spindle

(1) Contact sealings (see Figure 3-16) They are usually applied on clean working

environment.

（2）Non-contact sealings（see Figure 3-17） If the work environment fills with chip or dust, scraping between the cover and the shaft can be avoided by using non-contact sealings. They are effective for both oil and grease sealings.

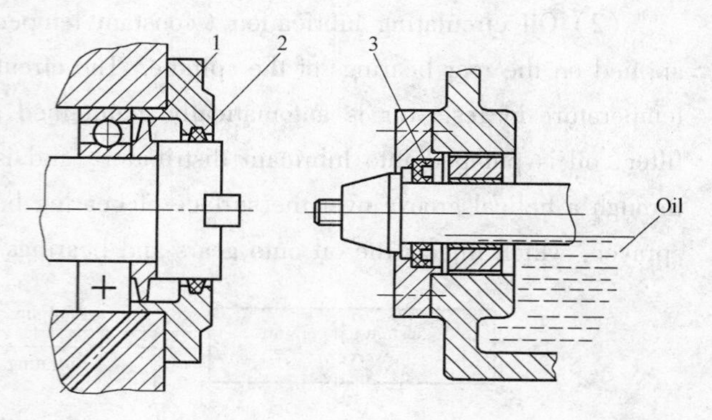

Figure 3-16 Contact sealings on shaft

1—Oil slinger 2—Felt ring 3—Rubber sealing

主轴的密封分接触式（图 3-16）和非接触式两种。

接触式利用轴承盖与轴的间隙进行密封，轴承盖孔内开槽则是为了提高密封效果。这种密封形式用在工作环境比较清洁的油脂润滑处。

非接触式密封分为环形槽密封、甩油环密封、迷宫式密封（图 3-17），甩油环密封是在螺母的外圆上开锯齿形环槽，当油向外流时，主轴传动的离心力把油沿斜面甩到端盖的空腔内，油液流回箱内；迷宫式密封结构在切屑多、灰尘大的工作环境下可获得可靠的密封效果，这种结构适用于油脂或油液润滑的密封。

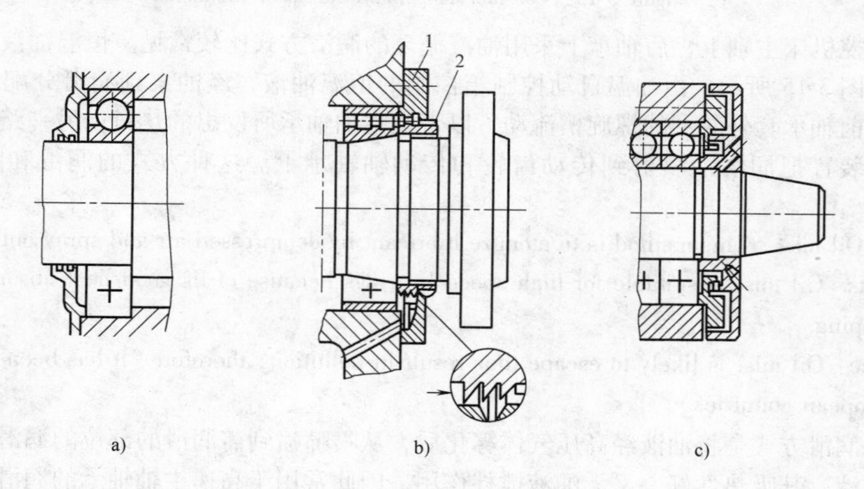

Figure 3-17 Non-contact sealings on shaft

a）Annular grooves b）Oil slinger c）Labyrinth

3.2 Feeding system

3.2.1 The requirements of feeding system

Feeding drive system is defined as the entire mechanical transmission chain that converts servo motor rotation into worktable linear motion. To ensure transmission precision and stability, feeding system must meet these requirements:

① Low friction. The friction between components tends to cause creeping during low speed motion. Ball screw and rolling guideways are widely applied on CNC machine tools to reduce friction between moving components. Static pressure guideways have ultra-low friction, but cost much.

② High transmission rigidity. Transmission rigidity depends on the rigidity of lead screw, worm-gear and their support framework. Cross-section and dimension of components must be carefully designed. Preload is usually applied.

③ Reduced moment of inertia. Besides satisfying strength and rigidity, minimized diameter and mass for each component should be pursued.

④ High resonance frequency. Inherent frequency of transmission system should be 2-3 times higher than work frequency of the servo system to avoid sympathetic vibration. Damper is recommended.

⑤ Minimized transmission clearance. Clearance eliminating design should be applied on each connection of transmission, e. g. gear-gear, lead screw-nut, couplings, etc.

数控机床的进给系统是指将伺服电动机的旋转运动变为工作台直线运动的整个机械传动链。为确保数控机床进给系统的传动精度和工作平稳性等，数控机床进给传动系统必须满足如下要求：

① 摩擦阻力要小。在进给系统中要尽量减少传动件之间的摩擦阻力，以消除低速进给爬行现象，从而提高整个伺服进给系统的稳定性。

② 传动刚度要高。进给传动系统的传动刚度主要取决于丝杠副（直线运动）或蜗杆副（回转运动）及其支承部件的刚度。传动刚度不足会影响传动准确性。

③ 转动惯量要小。在满足传动强度和刚度的前提下，应尽可能使各零件的结构、配置合理，减小旋转零件的直径和质量，以减少运动部件尤其是高速运转部件的转动惯量。

④ 谐振频率要高。为了提高进给系统的抗振性，应使机械构件具有高的固有频率和合适的阻尼。一般要求机械传动系统的固有频率应高于伺服驱动系统的 2～3 倍。

⑤ 传动间隙要小。对传动链的各个环节，包括齿轮副、丝杠螺母副、联轴器及其支承部件等均应采用消除间隙的结构或措施。

3.2.2 Gear drive

1. Clearance eliminating for spur gears

(1) Eccentric sleeve adjusting Rotating eccentric sleeve about the axis of the motor can adjust the vertical poison of the motor with the pinion on the shaft. The relative vertical motion between the pinion and the static driven gear changes their center distance thus eliminating clearance, as shown in Figure 3-18.

偏心套式消除间隙结构通过转动偏心套，使电动机中心轴线与从动齿轮轴线在垂直方向上发生相对移动，减小两啮合齿轮的中心距，从而消除齿侧间隙，如图 3-18 所示。

(2) Shim axial-adjusting (for bevel gears) Pitch diameter varies with axial distance. Teeth clearance between the pinion and the gear can be eliminated by their relative axial movement due to shim thickness adjusting, as shown in Figure 3-19.

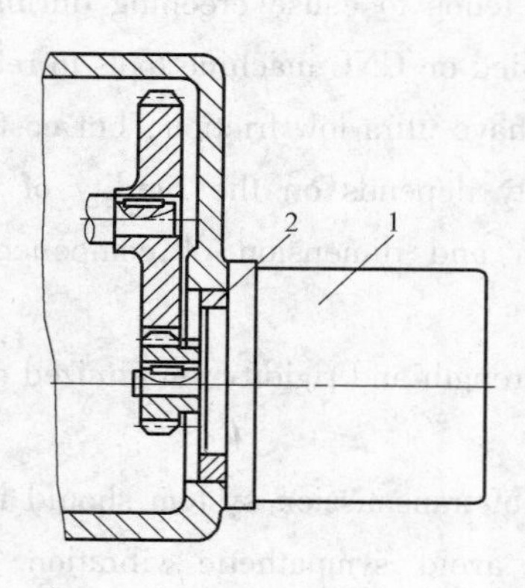

Figure 3-18 Eccentric sleeve adjusting mechanism

1—Motor 2—Eccentric sleeve

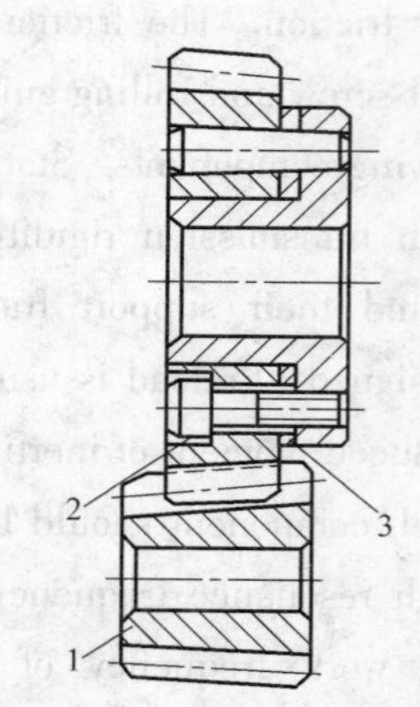

Figure 3-19 Shim axial-adjusting mechanism for spur gears

1—Pinion 2—Gear 3—Shim

由锥齿轮分度圆齿厚沿轴向方向略有锥度，可调整垫片厚度使两齿轮发生相对轴向移动，从而消除两齿轮的齿侧间隙，如图 3-19 所示。

(3) Double-gear staggered teeth adjusting (see Figure 3-20) The gear consists of the same halves (1, 2) that axially connected. The two halves can slightly perform relative rotation. After gearing with the pinion, tense the spring (4) by the nut (5) and lock the elongation by the nut (6). Due to reversion force of the spring, pins (3, 8) tend to rotate the half gears (1, 2) in opposite directions. As a result, each half has a teeth surface in one direction engages with the teeth surfaces of the pinion in both directions. Therefore, there is always a pair of teeth engaged without clearance in both rotational direction.

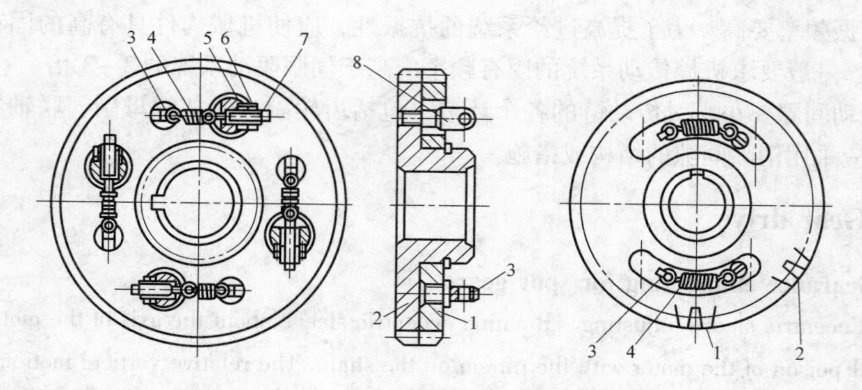

Figure 3-20 Double-gear staggered teeth adjusting mechanism

1、2—Thin gears 3、8—Pins 4—Spring 5、6—Nuts 7—Screw

双片薄齿轮错齿间隙消除结构如图 3-20 所示。两个齿数相同、同轴且相邻的薄片齿轮 1 和 2 与另一个宽齿轮啮合，装配后，通过螺母 5 调节弹簧 4 的拉力，并用螺母 6 锁紧。弹簧的拉力使薄片齿轮 1、2 具有相对转动的趋势，从而使两个薄片齿轮的左、右齿面分别贴在宽齿轮齿槽的左、右齿面上，实现齿侧间隙的消除。

2. Clearance eliminating for helical gears

The principle is similar to that of the double-gear staggered teeth adjusting, but the clearance eliminating is realized by axial movements of the halves to stagger the helical lines rather than relative rotation between the halves.

斜齿轮传动副消除间隙的方法与直齿轮双片薄齿轮消除间隙的原理相似，都是通过不同方法使两个薄片齿轮沿轴向移动，错开螺旋线。

Helical gear clearance eliminating can be realized by two methods:

(1) Shim axial-adjusting (see Figure 3-21) The thin gear halves (1, 2) were fixed together during manufacturing to give them exactly the same dimensions. During assembling, the shim (3) axially separated the halves to stagger their helical lines. Therefore, the halves engage with the pinion in opposite directions respectively.

However, the thickness of the shim must be precisely grinded. The mechanism is unable to compensate teeth wearing.

斜齿轮轴向垫片调整法的原理与错齿调整法相同。薄片斜齿轮1和2形状尺寸完全相同，装配时在两薄片齿轮间装入垫片3以错开螺旋线，使两薄片斜齿轮分别与宽齿轮4的左、右齿面贴紧，从而消除间隙，如图3-21所示。

垫片厚度必须经过精密的修磨才能调整好。这种结构的齿轮承载能力较小，且不能自动补偿间隙。

(2) Spring axial-compression (see Figure 3-22) The spring (3) produces compressive force to give the halves a relative axial movement thus making the halves engage with the pinion in opposite directions.

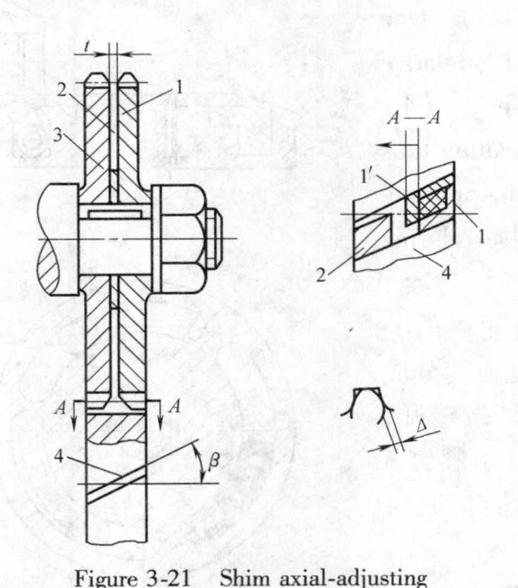

Figure 3-21 Shim axial-adjusting mechanism for helical gears

1、2—Helical gear halves 3—Shim 4—Pinion

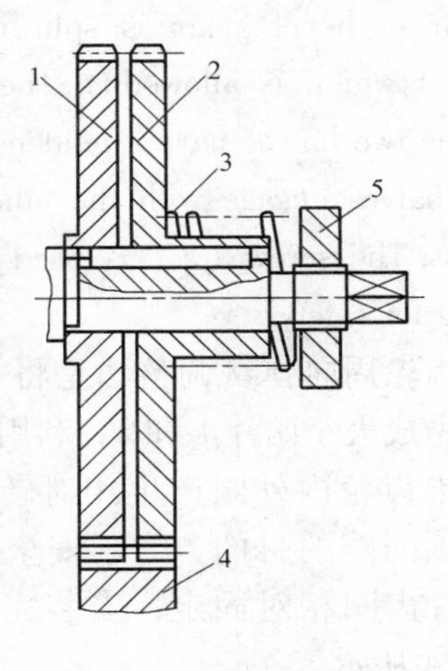

Figure 3-22 Axial spring compression for helical gears

1、2—Thin gear halves 3—Spring 4—Pinion 5—Nut

Compressive force of the spring can be adjusted by the nut (5). Insufficient compression cannot perform effective clearance eliminating, while excessive compression may shorten service life of the gearing. The mechanism can automatically compensate teeth wearing. Large axial dimension is the shortage of the system.

斜齿轮轴向压簧错齿调整结构的消隙原理与轴向垫片调整法相似，不同的是前者是利用薄片斜齿轮 2 右面的弹簧压力使两个薄片齿轮产生相对轴向位移，从而使它们的左、右齿轮面分别与宽齿轮的左、右齿面贴紧，以消除齿侧间隙，如图 3-22 所示。

弹簧 3 的压力可利用螺母 5 来调整，压力的大小要合适，压力过大会加快齿轮磨损，压力过小则达不到消隙作用。这种结构能使齿轮间隙自动消除，并始终保持无间隙的啮合。其缺点是轴向尺寸较大。

3. Clearance eliminating for bevel gears

The general principle of clearance eliminating of bevel gears is similar to that of the spur gears and helical gears.

(1) Axial spring compression (see Figure 3-23) One of the bevel gears is pushed by a compressed spring (3) that mounted on the shaft and can perform a slight axial movement to ensure a gearing without clearance. Compressive force of the spring is adjusted by the nut (4).

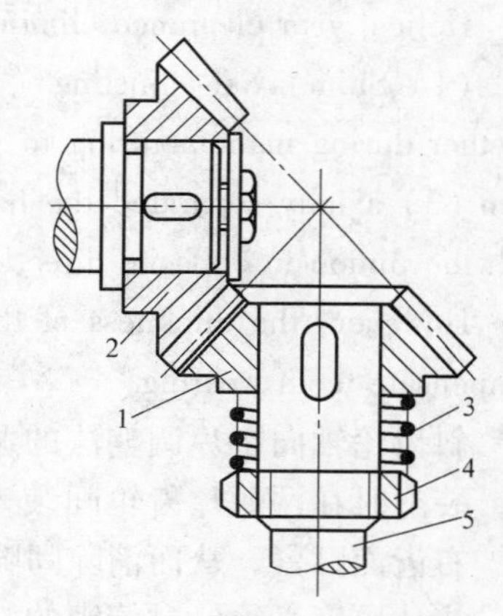

Figure 3-23 Axial spring compression for bevel gears
1、2—Bevel gears
3—Spring 4—Nut 5—Shaft

锥齿轮轴向压簧间隙消除结构中，锥齿轮 1 的传动轴上装有压簧 3，锥齿轮 1 在弹簧力的作用下可稍作轴向移动以消除间隙。弹簧力的大小由螺母 4 调节，如图3-23 所示。

(2) Tangential spring adjusting (see Figure 3-24) One of a pair of bevel gears is split into two halves (relative rotation of which is allowed). The compressed spring (4) makes the two halves have a tendency of relative rotating thus the two halves engage with the other bevel gear in opposite directions. The screw (5) is used to fix the halves during mounting for safety.

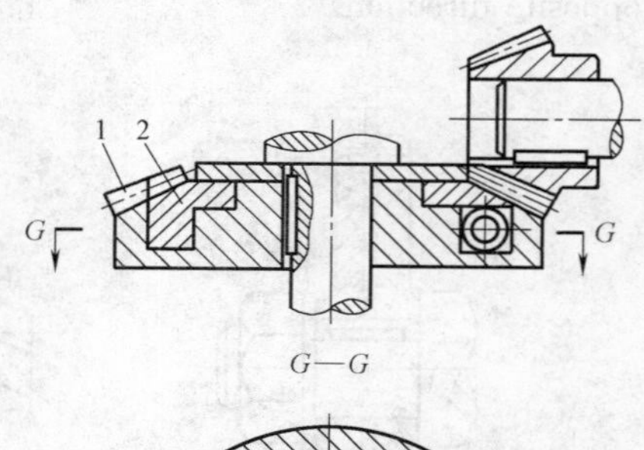

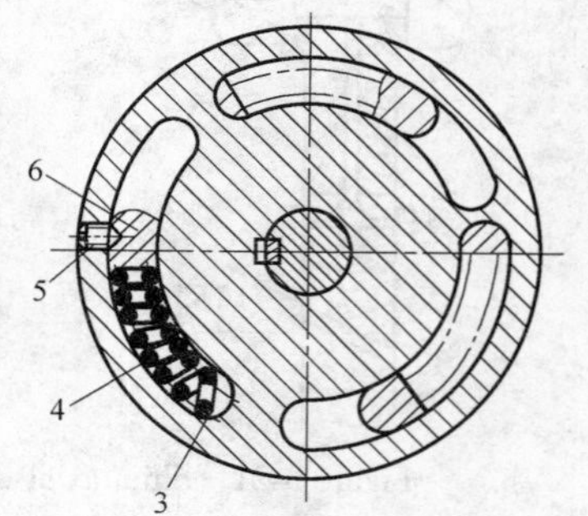

Figure 3-24 Tangential spring tension for bevel gears
1、2—Bevel gears 3—Paneling (镶块)
4—Spring 5—Screw 6—Claw

锥齿轮周向弹簧调整法是将一对啮合锥齿轮中的一个齿轮做成大小两片 1 和 2，利用弹簧 4 的回复力使分成两片的锥齿轮内外圈产生相对转动的趋势而稍微错开，达到消除间隙的目的。为了安全起见，安装时用螺钉 5 将大小片齿圈相对固定，安装完毕之后再将螺钉卸去，如图 3-24 所示。

4. Clearance eliminating for rack and pinion

Worktable stroke of a large CNC machine tool is too long to be driven by ball screw-nut system. Therefore, rack and pinion transmission is usually applied. For an accurate traveling, feeding, clearance between rack and pinion must be

eliminated.

In the system that shown in Figure 3-25, the shaft (2) is the driving shaft, mounted on which the gears has opposite helical lines. Apply an axial load on the shaft (2) to make the gears on it perform a slight axial motion. As a result, the shafts (1, 3) rotate in opposite directions to eliminate the clearances between the pinion (4) and the rack, and between the pinion (5) and the rack. When driving the rack in either direction, there is always a pair of engagement without teeth clearance between the rack and the pinion.

大型数控机床工作台的行程很长，进给运动不宜采用滚珠丝杠副来实现，而常采用齿轮齿条副来实现。

图 3-25 所示为齿轮齿条副间隙消除结构。进给运动由轴 2 输入，通过两对斜齿轮将运动传给轴 1 和 3，然后由两个直齿轮 4 和 5 驱动齿条，带动工作台移动。轴 2 上两个斜齿轮的螺旋线方向相反。如果通过弹簧在轴 2 上作用一个轴向力 ***F***，则使斜齿轮产生微量的轴向移动，这时轴 1 和轴 3 便以相反的方向转过微小的角度，使齿轮 4 和 5 分别与齿条的两齿面贴紧，从而消除间隙。

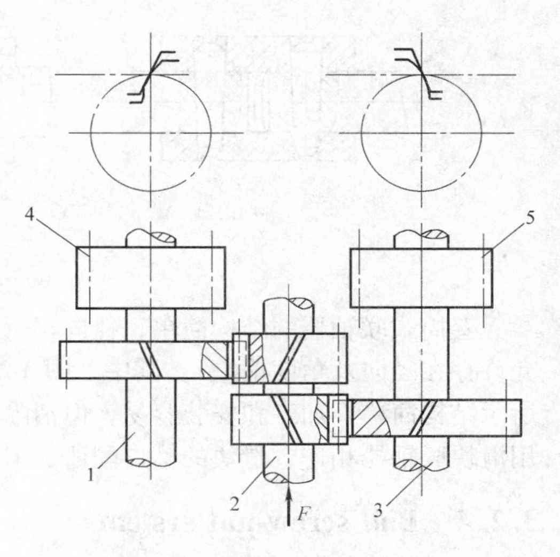

Figure 3-25 Clearance eliminating for the rack and pinion

1、2、3—Shafts 4、5—Pinions

3.2.3 Couplings

1. Conical ring keyless coupling

Different to conventional couplings, aconical ring keyless coupling delivers torque by the friction between the conical rings and the bush. When fastening the screw (5), conical rings (2) are axially compressed and have a tendency of radial expansion. The expansion is blocked by the bush (1) therefore producing friction. Maximum torque can be delivered depending on the number of conical rings pairs, as shown in Figure 3-26.

This configuration eliminates transmission clearance. Flexible conical rings can also accommodate slight concentricity between the shafts.

锥环无键联轴器机构利用锥环之间的摩擦实现轴与毂之间的无间隙连接传递转矩，并可调节两连接件之间的角度位置。通过选择所用锥环的对数，可传递不同大小的转矩。螺钉 5 通过压圈 3 施加轴向力时，由于锥环之间的楔紧作用，内外环分别产生径向弹性变形，消除配合间隙，同时产生接触压力以传递转矩，如图 3-26 所示。

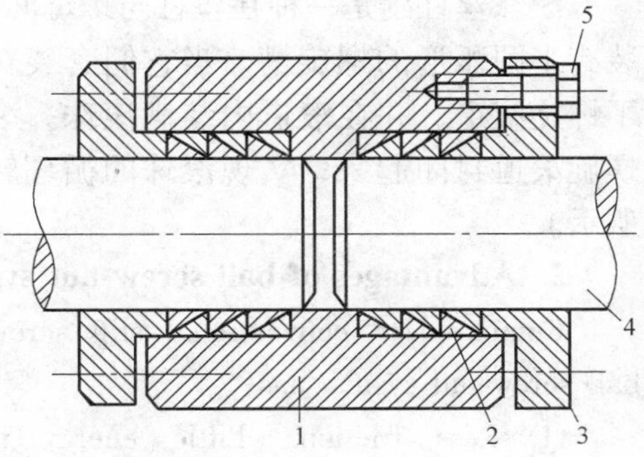

Figure 3-26 Conical ring keyless coupling connection

1—Bush 2—Conical rings 3—Cover 4—Driven shaft 5—Screw

2. Sleeve couplings

Sleeve couplings have simple constructions and

small radial dimensions, but they are unsuitable for precise mounting.

Type a: Pin connection (see Figure 3-27a). It has the simplest configuration, but effective shaft cross-sectional area is reduced;

Type b: Key connection (see Figure 3-27b). It is convenient for machining and installation, but tangential clearance between the key and the shaft is likely to exist;

Type c: Oldham coupling (see Figure 3-27c). The coupling consists of three parts, between which the sliding always exists thus transmission clearance eliminating is impossible.

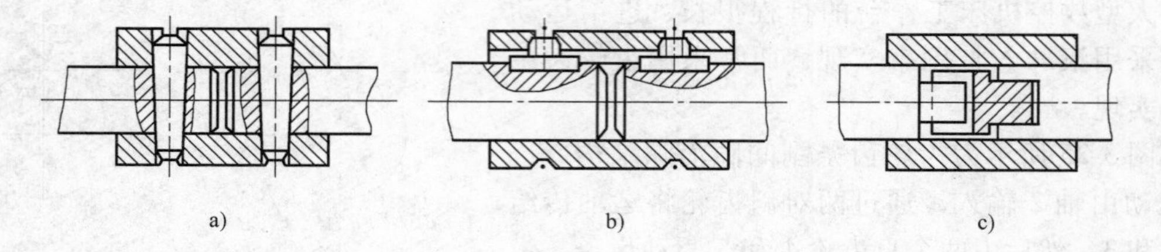

Figure 3-27 Sleeve couplings

套筒式联轴器的结构简单，径向尺寸小，但拆装困难（要求两中心轴线严格对准，不允许存在径向及角度偏差）。其中，图3-27a所示结构虽然简单实用，但不太可靠；图3-27b所示结构简单，加工和安装容易，但消除周向间隙不可靠，且易松动；图3-27c所示结构是用滑块联轴器相连，滑块的槽口配研，但这种结构无法消除传动间隙。

3.2.4 Ball screw-nut system

1. Working principle

Steel balls are installed between the screw and the nut to reduce the friction during rotation. Helical grooves of half-circle cross-section are machined both on the outer cylindrical surface of the screw and inner surface of the nut. The grooves with a return pipe compose a closed pipe to retain the steel balls for circulation. When the system is working, balls rotate about their own axes, and also rotate along the axis of the pipe, as shown in Figure 3-28.

滚珠丝杠副是一种在丝杠与螺母间装有滚珠作为中间元件的丝杠副。在丝杠和螺母上都装有半圆弧形的螺旋槽，当它们套装在一起时便形成了螺旋滚道，并在滚道内装满滚珠。滚珠回路管道将螺旋滚道封闭起来，实现滚珠的循环转动，如图3-28所示。

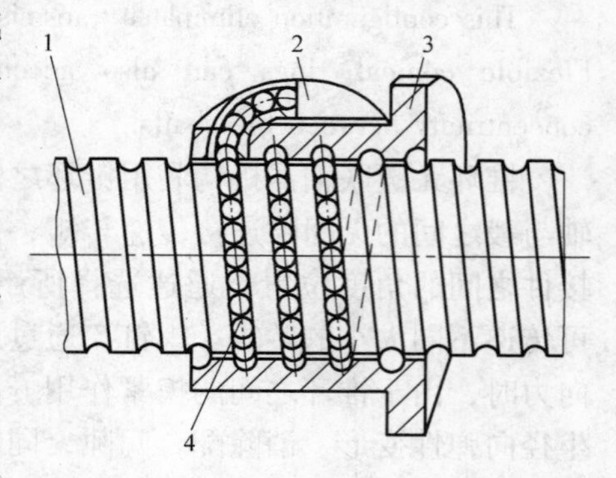

Figure 3-28 Construction of ball screw-nut system

1—Ball screw 2—Pipe 3—Nut 4—Balls

2. Advantages of ball screw-nut system

Compared to conventional slip screw-nut system, a ball screw-nut system has:

① Less friction. Little energy is consumed to overcome friction, transmission efficiency therefore has been increased by 3- 4 times ($\eta = 0.92\text{-}0.96$). Less friction also results in less heat emission thus enables ultra high rpm;

② Stable transmission. Because of the ultra low friction, creeping is not likely to occur;

③ High transmission accuracy and rigidity. By preloading, transmission rigidity has been improved, and reverse clearance can be eliminated;

④ Long service life. Low friction results in a low wearing rate, therefore, the system can maintain its accurate operation for a long term;

⑤ Motion reversibility. Low friction invalidates self-locking, therefore, rotational motion and linear motion can be converted, i. e. either the screw or the nut can be the driving or driven component in the system.

However, the complex construction of the system requires high quality of components thus high cost. For vertical transmission, breaking and balance system must be considered since self-locking cannot be applied.

与传统的滑动丝杠副比较，滚珠丝杠副具有以下优点：

① 滚珠丝杠副的传动效率高达 0.92 ~ 0.96，是普通丝杠副的 3 ~ 4 倍，能耗低，发热少，适合高速运动。

② 摩擦阻力小，动、静摩擦系数之差极小，因此运动平稳，不易出现爬行现象。

③ 滚珠丝杠副经预紧后，可消除轴向间隙，因而传动精度高，反向无死区。

④ 磨损小，精度保持性好，使用寿命长。

3. Types of ball screw-nut units (by construction)

(1) External circulation In an external circulation (see Figure 3-29), balls return to the start point through a pipe that connects both ends of the helical groove.

External circulation has the most industrial application due to its simple construction, good manufacturability and heavy loadability (for heavy transmission). Smaller radial dimension is desired.

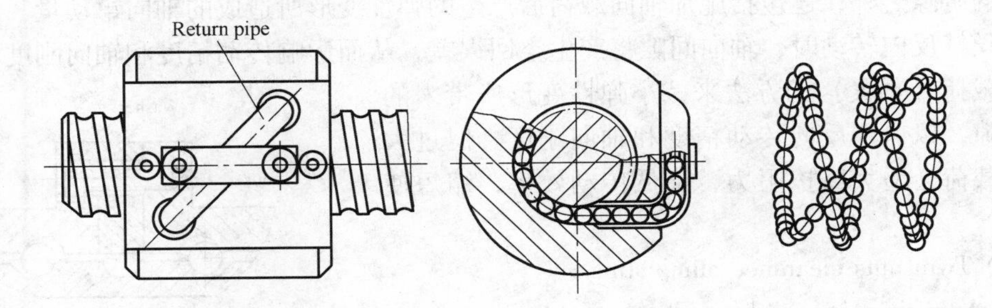

Figure 3-29 External circulation of ball screw-nut system

滚珠到达丝杠螺母螺旋槽末端后，通过螺母外表面上的插管返回到螺旋槽起点重新进入循环的工作方式称为外循环，如图 3-29 所示。因为结构简单，工艺性好，承载能力较高，目前采用这种循环方式的滚珠丝杠副应用最为广泛，但其缺点是径向尺寸较大。

(2) Internal circulation In an internal circulation (see Figure 3-30), each thread of the nut has a closed circulation, and the balls are rotating only in the same thread. The number of return pipes equals to that of threads.

Internal circulation has a compact construction and high rigidity. Reduced system friction provides a fluent traveling of balls. It is suitable for high sensitive and accurate feeding system. Compared to external circulation, it has less effective number of balls in each thread. Therefore, internal circulation is not designed for heavy duty.

内循环式滚珠丝杠通过螺母上安装的反向器接通相邻滚道，使滚珠单圈循环，如图 3-30 所示。这种类型的循环结构紧凑，刚度好，滚珠流通性好，摩擦损失小，适用于高灵敏、高精度的进给系统，但制造较困难，且不宜用于重载传动。

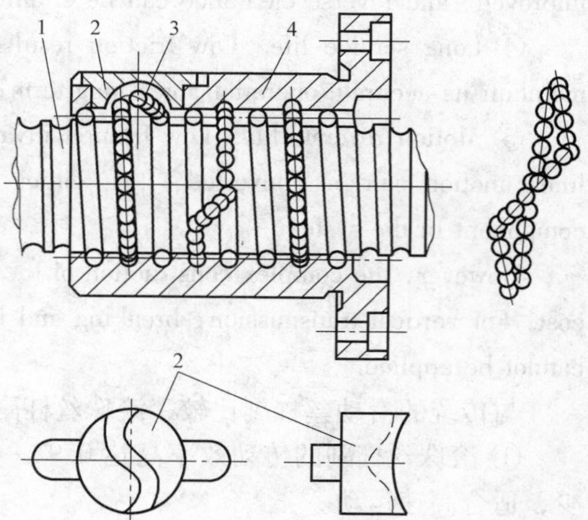

Figure 3-30 Internal circulation of ball screw-nut system

1—Screw 2—Return pipe 3—Balls 4—Nut

4. Clearance adjusting of ball screw-nut assembly

Axial clearance is defined as unexpected axial motion between the ball screw and the nut without relative rotation. Axial clearance is caused by manufacturing errors and elastic deformation of system components under loading.

When the screw rotates reversely, the actual displacement is less than that is expected due to axial clearance. That is unfavorable for transmission accuracy and rigidity. Therefore, preloading is necessary to applied to eliminate the axial clearance. The preloading force must be accurately adjusted, otherwise, excessive preloading increases friction, thus decreases transmission efficiency, and shortens the service life of the system.

滚珠丝杠的轴向间隙是指丝杠和螺母无相对转动时两者之间的最大轴向窜动量，除了结构本身的游隙之外，还包括施加轴向载荷后产生的弹性变形所造成的轴向窜动量。

当丝杠反向转动时，轴向间隙将产生空回误差，从而影响传动精度和轴向刚度。通常采用预加载荷（预紧）的方法来减小弹性变形所带来的轴向间隙，以保证反向传动精度和轴向刚度。但过大的预加载荷会增大摩擦阻力，降低传动效率，缩短使用寿命。

(1) Twin nuts clearance eliminating

① Adjusting by washer. Washer adjusting (see Figure 3-31) provides the simplest method of clearance eliminating. Relative axial position of the nuts is determined by the thickness of the washer between the nuts.

Washer adjusting provides enough transmission rigidity. However, adjustment is inconvenient, and clearance eliminating cannot be retained for a long term due to track wearing.

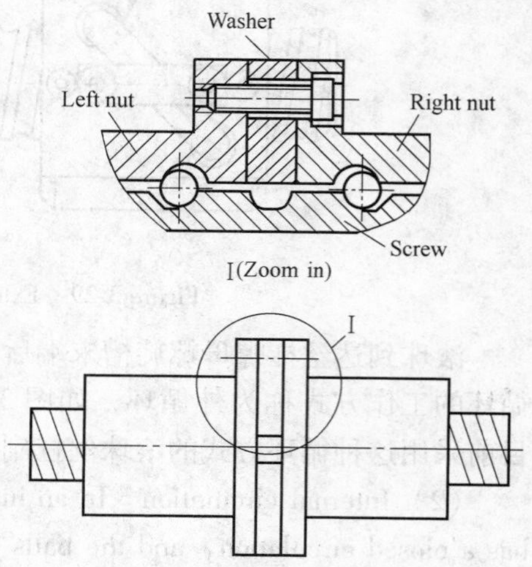

Figure 3-31 Clearance eliminating by washer adjusting

垫片调隙式通过调整垫片厚度，使左、右两螺母产生相对轴向位移来消除间隙，并产生预紧力，如图 3-31 所示。这种方法结构简单，刚性好，但调整不便，滚道有磨损时不能随时消除间隙和进行预紧。

② Adjusting by thread. One of the ball nuts is fixed to the machine tool by flange, and the other has thread on one end, onto which the adjusting nut and locking nut are attached (see Figure 3-32). Turning the adjusting nut causes axial motion of the ball nut to eliminate clearance and produce preload.

Figure 3-32 Clearance eliminating by nut adjusting
1—Adjusting nut 2—Locking nut

The method is simple and convenient, but preloading force cannot be precisely controlled.

在螺纹调隙机构中，一个滚珠螺母通过外端凸缘与机床固定，另一个滚珠螺母的外端制有螺纹，伸出套筒外，并附有两个圆螺母，如图 3-32 所示。旋转圆螺母时，即可消除轴向间隙，并产生预紧力。调整好后再用另一个圆螺母锁紧。这种方法调整方便，但难以准确控制预紧力的大小。

③ Adjusting by tooth difference. Spur gears are machined on the flanges of the ball nuts. Tooth numbers are z_1 and z_2, and $z_2 - z_1 = 1$ (see Figure 3-33).

Rotate the nuts for an angle corresponding one tooth on their flanges to produce a relative angular motion between the nut, then the controllable minimum axial displacement between the nuts is:

$$s = \left(\frac{1}{z_1} - \frac{1}{z_2}\right)P_n$$

Where P_n is pitch of the screw bar.

This system provides precise preloading, not only convenient but also reliable. It is usually applied on where high accuracy is required.

在齿差调隙装置中，两个螺母的凸缘上均制有圆柱齿轮，分别与紧固在套筒两端的内齿圈相啮合，其齿数分别为 z_1 和 z_2，并相差一个齿，如图 3-33 所示。若将两个螺母相对于套筒方向都转动一个齿，则两个螺母产生相对转动角位移，其最小可控轴向位移量为 $s = \left(\frac{1}{z_1} - \frac{1}{z_2}\right)P_n$。这种调整方法能精确调整预紧量，调整方便、可靠，但结构尺寸较大，多用于高精度的传动。

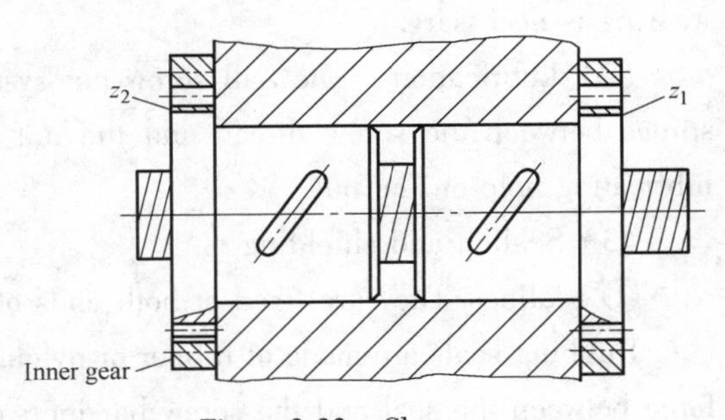

Figure 3-33 Clearance eliminating by tooth difference

(2) Single nut eliminating

① Deflected thread. At the middle of the nut length, one of the threads has a sudden pitch variation of ΔL_0, therefore, the balls near the deflected thread are forced to contact the grooves.

This method is simple and compact, and is especially suitable for the small size ball screw-nut system. However, the preloading force depends on the amount of pitch deflection, which cannot be adjusted once the nut is designed (see Figure 3-34).

变位螺距预紧法的原理是在螺母轴向中部的一圈螺纹上产生一个轴向的导程突变量，从而使左右端的滚珠在轴向错位实现预紧。这种调隙方法结构简单紧凑，运动平稳，特别适用于小型丝杠副，但预紧力须预先设定，且不能随意改变，如图 3-34 所示。

② Preloading by set screw. When the manufacturing of the nut is completed, a radial slot will be cut on the nut. Fastening the adjusting screw actually shrinks inner diameter of the nut, therefore, preloading force is applied onto the balls between the helical grooves (see Figure 3-35).

This method has good cost effectiveness, and is especially convenient for preloading and clearance adjusting. The solution of balls' flowability near the slot is a patent.

螺母在专业生产工厂完成精磨之后，沿径向开一薄槽，通过调整螺钉实现间隙的调整和预紧，如图 3-35 所示。该专利技术已成功地解决了开槽后滚珠在螺母中的通过性问题。单螺母螺钉结构不仅具有很好的性价比，而且间隙的调整和预紧极为方便。

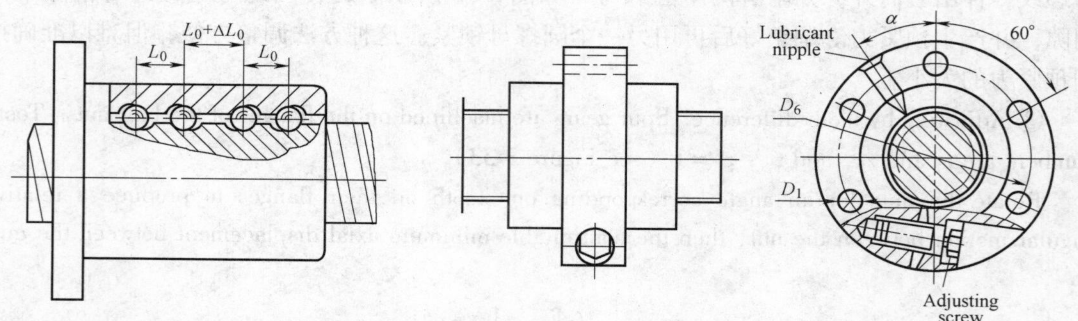

Figure 3-34 Clearance eliminating by pitch deflection

Figure 3-35 Screw preloading

5. Maintenance of the ball screw-nut system

(1) Regular examination Regular check of the connections and bearings for their fastness and wearing is necessary.

(2) Lubrication The ball screw-nut system must be regularly lubricated. Grease should be spread between the screw thread and the nut housing. Oil is supplied into the system through a lubricating hole on the nut.

(3) Sealing and shielding

① Sealings. They are fixed at both ends of the nut to protect the balls.

Rubbing seals are made of rubber or nylon. It engages with the thread of the screw bar. Contact force between the seal and the screw bar increases friction.

Non-rubbing seals is known as labyrinth rings, and usually made of PVC. A clearance exists between the inner surface of the sealing and the thread of the screw bar thus no friction.

② Shields. Helix steel band and flex sleeve shields (see Figure 3-36) are commonly used. One end of a shield is fixed onto the nut and teh other end on the screw support. Shields effectively protect the thread from chippings and dust.

Figure 3-36 Flex sleeve shields

6. Brake system of ball screw-nut

Ultra low friction makes self-lock impossible for a ball screw-nut system. Therefore, the brake must be applied for vertical motion components to prevent sliding downwards due to gravity. Figure 3-37 shows the brake system of a CNC milling machine.

由于滚珠丝杠副的摩擦系数极小，无自锁作用，所以当滚珠丝杠副垂直传动时，为防止因自重下降，必须装有制动装置。

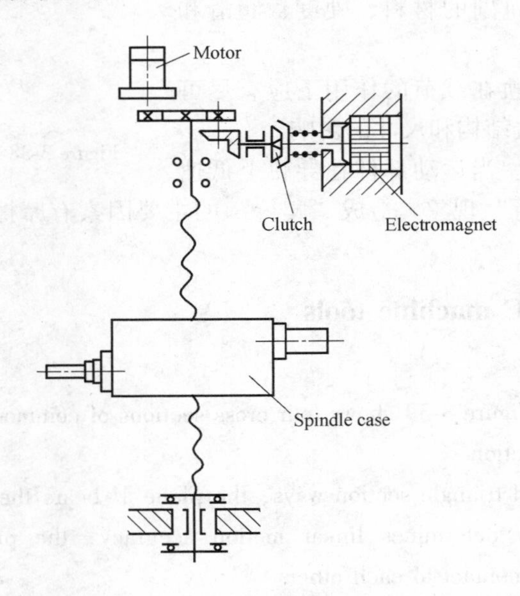

Figure 3-37 Spindle braking system of a CNC milling machine

3.3 Ways

3.3.1 The requirements of ways

The ways support machine tool components, and lead them to perform linear or circular motions (see Figure 3-38).

To ensure accuracy and stability of the machine tool, the ways must have:

① High guiding accuracy. It depends on shape, construction, manufacturing accuracy and clearance adjusting of the ways.

② High abrasion resistance. It enables the ways to remain guiding accuracy for a long time. Abrasion resistance depends on material, hardness, lubrication and load of the ways.

③ High rigidity. Distortion-resistance increases with rigidity. Rigidity is affected by construction and dimensions of the ways.

④ Stable low speed motion. Creeping at low speed should be avoided. Creeping is caused by friction characteristics, lubrication conditions and transmission rigidity.

导轨用于支承机床部件，并引导其作直线或圆周运动，如图 3-38 所示。为确保机床的精度与稳定性，导轨必须满足以下要求：

Figure 3-38　A pair of way and slide

① 导向精度高。导向精度保证部件运动轨迹的准确性，它受导轨的结构形状、组合方式、制造精度和导轨间隙调整等的影响。

② 良好的耐磨性。耐磨性好可使导轨长久保持导向精度。耐磨性受导轨副的材料、硬度、润滑和载荷的影响。

③ 足够的刚度。导轨在载荷的作用下应有尽可能小的变形。刚度受导轨结构和尺寸的影响。

④ 低速运动平稳性。当运动部件在导轨上低速移动时，不应发生“爬行”现象。造成“爬行”的主要因素有摩擦性质、润滑条件和传动系统的刚度。

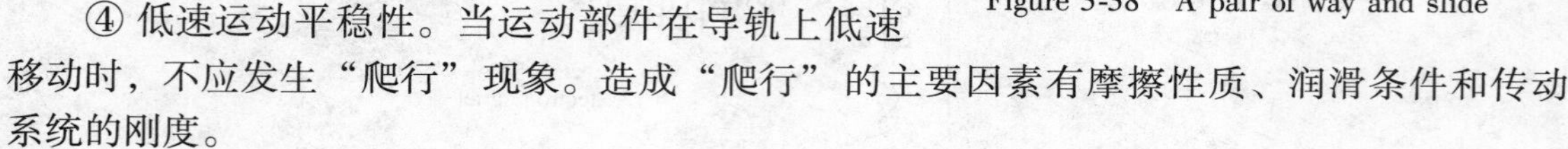

3.3.2　Ways for CNC machine tools

1. Slide ways

(1) Constructions　Figure 3-39 shows four cross sections of commonly used ways. Each plane on a way has a unique function.

For square-section and triangle-section ways, the plane *M* bears the weight of the components on the way; the plane *N* determines linear motion accuracy; the plane *J* forces the moving component and the way to contact to each other.

For dovetail-section ways, the plane *M* leads the moving component and keeps contact; the plane *J* bears the weight on the way.

滑动导轨常见的截面形状如图 3-39 所示。导轨上的不同平面所起的作用各不相同。在矩形和三角形导轨中，*M* 面主要起支承作用，*N* 面作导向面，*J* 面是防止运动部件抬起的压板面；在燕尾形导轨中，*M* 面起导向和压板作用，*J* 面起支承作用。

(2) Configurations of slide ways　A pair of triangle-sectionway and square-section way is popular in most CNC machine tools. Dovetail-section is seldom used.

Square-section ways guide moving components on the way by their sides. If both sides of one of the ways are used for guiding, it is known as narrow guiding (see Figure 3-40a). A pair of ways of narrow guiding is easy to manufacture, and has less influence by thermal distortion.

For broad guiding (see Figure 3-40b), each way contributes a side to guide moving

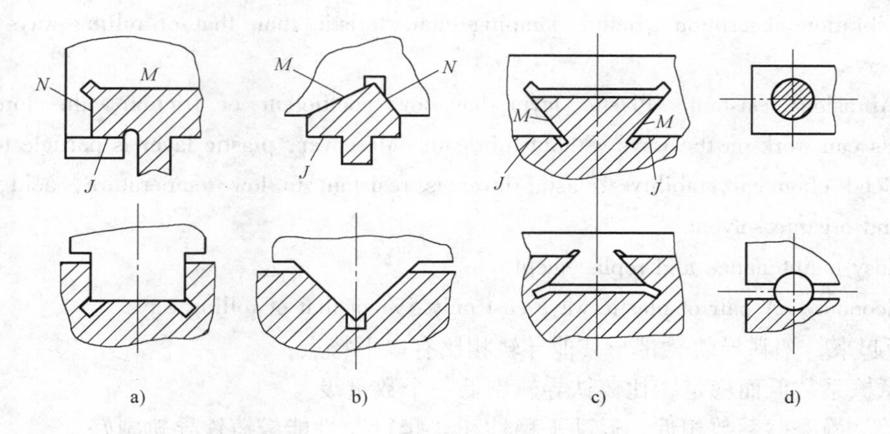

Figure 3-39 Cross sections of slide way pairs
a) Square b) Triangle c) dovetail d) Column

component. Compared to narrow guiding, wide guiding has better accuracy and stability.

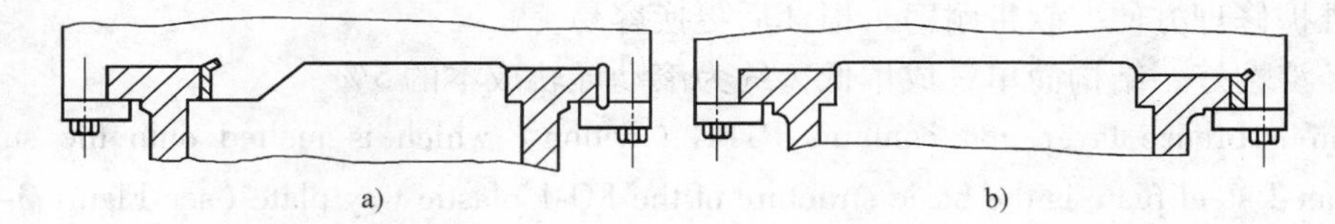

Figure 3-40 Configurations of ways
a) Narrow guiding b) Broad guiding

滑动导轨一般由两条导轨组成。数控机床滑动导轨的组合形式主要是三角形配矩形式和矩形配矩形式。

双矩形导轨是用侧边导向，当采用一条导轨的两侧边导向时称为窄式导向，窄式导向导轨制造容易，受热变形影响小。分别采用两条导轨的两个侧边导向则称为宽式导向。

Conventional ways are made of cast iron. They have large coefficient of static friction, and the coefficient of dynamic friction varies with motion speed. Motion stability and accurate positioning will be affected by creeping at low speed. Since conventional cast iron ways cannot meet the requirement of modern CNC machine tools, they have almost been replaced by plastic slide ways.

传统的铸铁导轨静摩擦系数大，动摩擦系数随速度而变化，摩擦损失大，低速时易出现“爬行”现象，影响运动平稳性和定位精度，因此目前在数控机床上已很少采用，且被塑料滑动导轨取代。

Plastic slide ways are actually not made of plastic, but with a plastic layer onto the surface of cast iron or steel ways. Recently, plastic way plates and soft plastic bands are commonly used as surface layers in China. They are made of composite material that consists of “Teflon” and certain metals.

Compared to conventional cast ironways, plastic slide ways have these advantages:

① Low coefficient of friction (10% of that of conventional ways).

② Similar coefficient of static friction and dynamic friction, better stability and less creeping.

③ Vibration absorption. Better damping characteristic than that of rolling ways and static pressure ways.

④ Abrasion resistant. Plastic layer has low coefficient of friction, therefore, moving components can work on the ways without lubricant. Moreover, plastic layer is particle tolerant.

⑤ Good chemical stability. Plastic layer is resistant to low temperature, acid, alkaline, oxidizer and organic solvent.

⑥ Easy maintenance and replacement.

⑦ Economy. A pair of plastic ways cost only 5% of that of rolling ways.

概括起来，塑料导轨软带与其他导轨相比有以下特点：

① 摩擦系数低而稳定，比铸铁导轨副低一个数量级。

② 动、静摩擦系数相近，运动平稳性和“爬行”性能较铸铁导轨副好。

③ 吸振性好，其阻尼特性优于接触刚度较低的滚动导轨和易漂浮的静压导轨。

④ 耐磨性好，有自身润滑作用，无润滑剂也能工作，灰尘磨粒的嵌入性好。

⑤ 化学稳定性好，耐低温，耐强酸、强碱、强氧化剂及各种有机溶剂。

⑥ 维护修理方便，软带耐磨，损坏后更换容易。

⑦ 经济性好，结构简单，成本低，约为滚动导轨成本的5%。

A porous bronze layer that contains PTFE (Teflon) which is melted onto the surface of a bronze-coated steel plate is the basic structure of the FQ-1 plastic way plate (see Figure 3-41). It is suitable for vertical mounting where lubricating is inconvenient since it has a coefficient of friction of only 0.04-0.08.

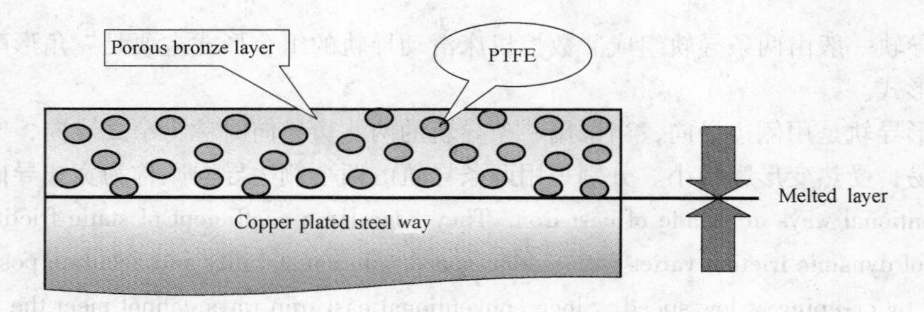

Figure 3-41 Construction of the FQ-1 plastic way

例如，FQ-1塑料导轨板采用的是在镀铜钢板上烧结一层多孔青铜，在青铜层间隙中扎入聚四氟乙烯及其他填料，再经适当处理形成金属－氟塑料的复合体导轨板，如图3-41所示。这种导轨板的摩擦系数仅为0.04～0.08，并具有良好的自润滑作用，特别适用于润滑不便的立式装备。

The soft plastic band is made of PTFE, which has been added with powder mixture of bronze, MoS_2, graphite, etc (see Figure 3-42). Under oil lubricating, coefficient of friction of the plastic band is approximately 0.06. It has 8-10 times the service life than a conventional cast iron way. Plastic bands in different thickness are available, and can be custom-tailored by users. Sealing is necessary to protect the band from scraping.

塑料导轨软带是以聚四氟乙烯PTFE为基体，添加青铜粉二硫化钼和石墨等多种填料所

构成的复合材料，如图3-42所示。在油润滑状态下，其摩擦系数约为0.06，使用寿命为普通铸铁导轨的8~10倍。塑料导轨软带有各种厚度的规格，长与宽由用户自行裁剪，采用粘贴的方法固定。塑料软带较软，易被硬物刮伤，因此要有良好的密封防护措施。

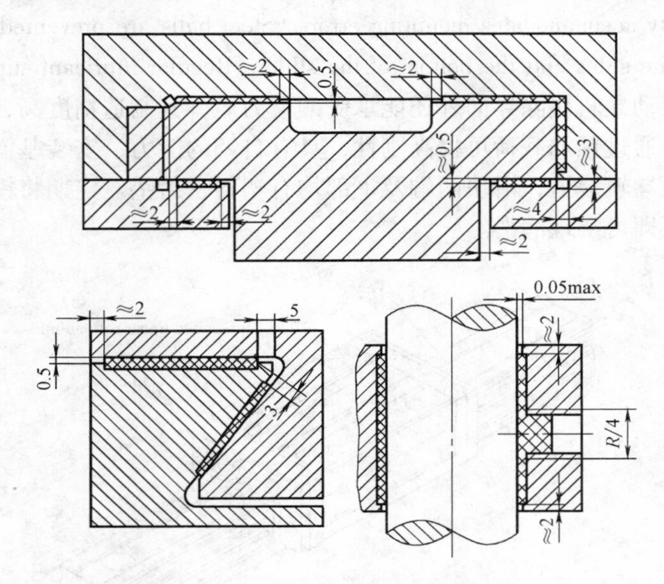

Figure 3-42 Mountings of soft bands on ways

2. Rolling ways

(1) Types

① Rolling unit (see Figure 3-43a). In each unit, rolling elements (usually steel rollers) perform circulation in a closed track. Rolling units are professionally manufactured. Different specifications are available for users. In a machine tool, moving component is mounted onto rolling units. Figure 3-43b shows the assembly of rolling units between the slide and the way. The number of rolling units being used depends on load and length of the ways, but at least two on each way. Ways are mostly quenched steel-embedded. Rolling units are suitable for medium load application.

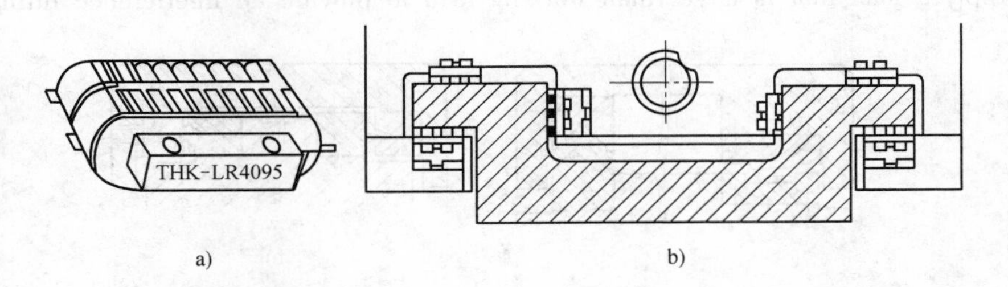

Figure 3-43 Rolling unit and system assembly

a) Rolling unit b) Slide assembly

滚动导轨块是一种以滚动体作循环运动的滚动体。移动部件移动时，滚动体沿封闭轨道作循环运动。滚动体为滚珠或滚柱。滚动导轨块由专业厂家生产，有多种规格、形式供用户选用，多用于中等负荷导轨。使用时，导轨块安装在运动部件上，导轨块的数目取决于导轨的长度和负载的大小，但每个导轨应至少用两块。

② Integrated linear rolling unit. Besides guiding, an integrated linear rolling unit (see Figure 3-44) can withstand turnover moment. During working, rollers perform circular rotation in a track between the way and the slider. The system has long term accuracy and allows high speed operation. Self-alingment ability accommodates mounting errors. Steel balls are prevented from dust and chips by sealing at both the sides and the bottom of the slider. Regular lubricant supply is required.

整体式直线滚动导轨块除导向外还能承受颠覆力矩，其制造精度高，可高速运行，并能长时间保持精度，通过预加载荷可提高刚性，具有自调的能力，安装基面允许误差大。为防止灰尘和脏物进入导轨滚道，滑块两端及下部均有塑料密封垫。定期将锂基润滑脂放入滑块上的润滑油注油杯即可实现润滑。

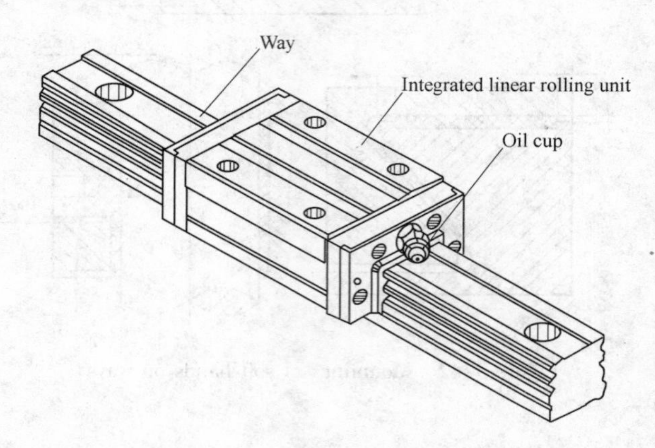

Figure 3-44 Application of integrated linear rolling unit

(2) Preloading of rolling ways (see Figure 3-45) The rolling unit must be preloaded to eliminate clearance between the balls and the track. Preloading may improve contact rigidity and transmission accuracy of ways. For vertical ways, preloading also prevents rollers from escaping, and keeps the rolling unit perpendicular.

Preload can be performed in two ways:

① Apply a load that is larger than working load to provide an interference fitting of 2-3μm

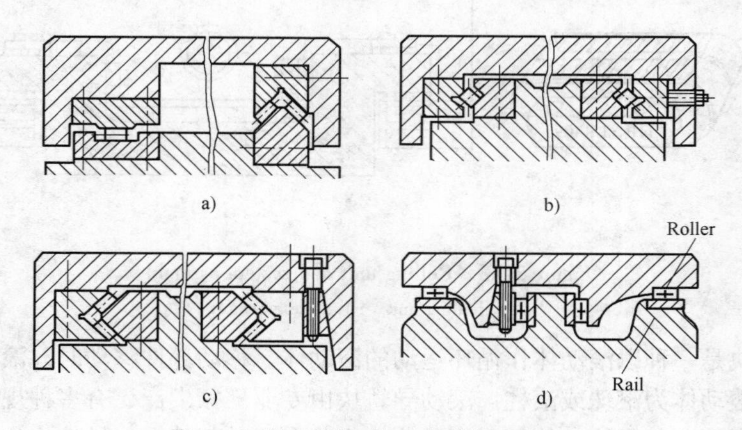

Figure 3-45 Preloading of rolling ways

a) Gravity preloading b) Set screw preloading c) & d) Wedge preloading

during mounting. Large interference results in over-preloading thus wasting driving energy. Heavy components may preload the system by their weight.

② After assembling, perform preload by adjusting screw, wedge, or eccentric wheel.

为了提高滚动导轨的刚度，应对滚动导轨进行预紧，以提高接触刚度，消除间隙，如图3-45所示。在立式滚动导轨上，预紧可防止滚动体脱落和歪斜。

3. Hydraulic ways

For a pair of static pressure ways, the space between the way pairs is filled with high pressure hydraulic oil to float the moving component (see Figure 3-46). Most hydraulic ways are equipped with oil supply system of constant pressure. During operating, oil pressure can be automatically adjusted according to load variation to keep moving component floating.

Hydraulic ways has ultra-low coefficient of friction (approximately 0.0005) to eliminate creeping and vibration during operating. Mechanical friction free between components also means no abrasion, therefore, long term accuracy and long service life can be realized.

However, hydraulic ways are expensive, because they have complex constructions, and require extra hydraulic systems. They are only applied on large size or heavy duty CNC machine tools.

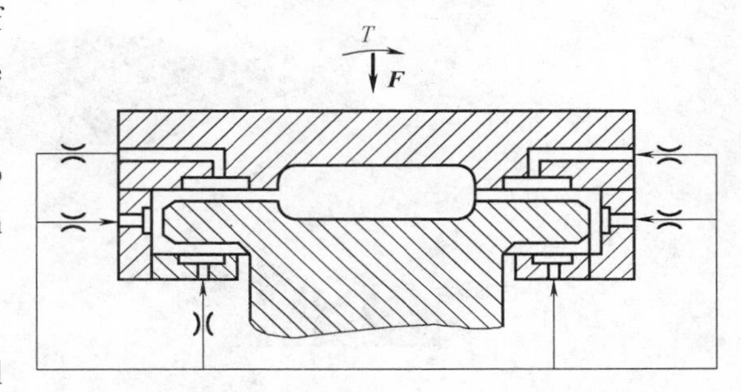

Figure 3-46　Work principle of hydraulic ways

液压导轨的原理是在两个相对运动的导轨面之间通入压力油，使运动件浮起，如图3-46所示。自动油压调节装置可保证导轨面始终处于纯液体摩擦状态。静压导轨的摩擦系数极小(约0.0005)，导轨运行平稳，既无“爬行”，也不会产生振动，功率消耗小，导轨不会磨损，有利于提高导轨的精度保持性和使用寿命。但液压导轨结构复杂，并需要一套专门的液压装置，因此制造成本较高，多应用在大型、重型数控机床上。

3.3.3　Lubrication and protection for ways

1. Lubrication

For low speed ways and seldom operated ways, regular manual lubricating and oil cup supply are preferred for economy, but lack of reliability. Ways of most CNC machine tools are under active lubricating, i.e. pumping lubricant to ways. Besides lubricating, it also performs cleaning and cooling for ways. The active lubrication system is a basic equipment of most standard CNC machine tools.

Good lubricant should have stable viscosity at a wide temperature range, and can form a rigid oil film to withstand forces during operation.

运动速度低、工作不频繁的滚动导轨可采用人工定期加油或油杯供油等最简单的润滑方式，这些润滑方式较经济但可靠性较低。因此大部分数控机床都备有专门的供油系统，以压力油强制润滑，同时利用油的流动冲洗和冷却导轨表面。

优质润滑油应具有良好的粘温特性、良好的润滑性能和足够的油膜刚度，油液应洁净且不浸蚀机件。

2. Protection of ways

Chips, particles or cutting fluid may result in unexpected abrasion and corrosion of ways, therefore, ways must be shielded (see Figure 3-47).

Shields must be kept at good conditions to provide reliable protection. For full-metal shields, regular manual lubricating is required at joints and connection.

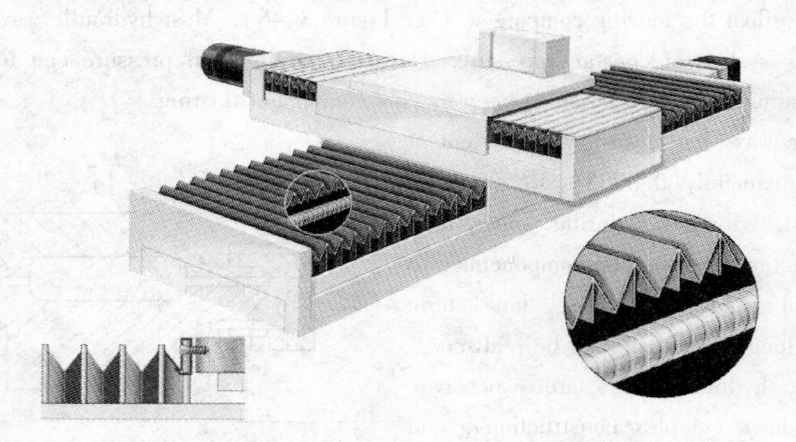

Figure 3-47　Foldable shields

为了防止切屑、磨粒或切削液散落在导轨面上而引起磨损、擦伤和锈蚀，导轨面上应有可靠的防护装置（图 3-47）。在机床使用过程中应防止损坏防护罩，金属防护罩的接缝和铰链处应经常用刷子蘸机油清理，以避免发生碰、卡现象。

3.4　Position sensors of CNC machine tools

The servomechanism is the driving system of a CNC machine tool. The servomechanism and the transmission system convert control signals of the computer to mechanical motions of machine tool components. As an important part of a servomechanism, position sensors monitor machine tool motion by comparing the actual position and motion speed to that of instruction. Compensation will be made to eliminate transmission errors.

Sensitivity of position sensors determines accuracy of the machine tool. Usually, measuring sensitivity is approximately ±（0.001-0.02）mm. A reasonable sensitivity is between 1/10-1/3 of machine tool accuracy.

伺服系统是机床的驱动部分，计算机输出的控制信息通过伺服系统和传动装置驱动机床运动。位置检测装置是数控机床伺服系统的重要组成部分，其作用是检测位移和速度，发送反馈信号，并与数控装置发出的指令信号比较，控制执行部件向着消除偏差的方向运动，直至偏差为零。不同的数控机床对检测元件和检测系统的精度要求、允许的最高移动速度都不相同。一般要求检测元件的测量精度为 ±（0.001 ~ 0.02）mm/m。应当合理选择系统分辨

率，一般按机床加工精度的1/10～1/3选取。

3.4.1 The requirements of a satisfactory position sensor

① High reliability, EMI-free, insensitive to humidity and temperature varying.

② Well protected from dust, oil mist, chips. Easy maintenance.

③ Accurate and rapid response.

④ High speed automatic dynamic measuring and processing.

⑤ A balance between low cost and long service life.

对数控机床位置检测装置的基本要求是：

① 高可靠性和高抗干扰性。

② 使用维护方便，有防护措施。

③ 能够满足精度和速度的要求。

④ 易于实现高速的动态测量、处理和自动化。

⑤ 成本低，寿命长。

3.4.2 Types of position detecting devices

1. Digital and analog measurement

Digital measuring devices numerize measuring results, and represent them in digits. It has simple construction that would not be influenced by electromagnetic interference. Output signals are electronic pulses, which is convenient to display and process. Measuring accuracy depends on the pulse frequency. Grating displacement sensor is a typical digital measuring device.

Analog measuring devices represent measuring results by continuous variables (e.g. phase, voltage, etc.) without conversion. In a CNC machine tool, analog measuring is used for small displacement measuring, where high accuracy can be achieved. Eddy transformers and induction synchronizers are typical analog measuring devices.

按测量值的表现方式，位置检测装置可分为数字式测量和模拟式测量。

数字式测量输出信号一般是电脉冲，可以直接送到数控装置进行比较、处理。数字式测量装置比较简单，抗干扰能力强，测量精度取决于测量单位，与量程基本无关。光栅位移测量装置是典型的检测装置。

模拟式测量用连续的变量（如相位变化、电压幅值变化）来表示测量结果，无需进行变换。在数控机床上，模拟式测量主要用于小量程的测量，可以达到很高的精度。常见的模拟式测量装置有旋转变压器、感应同步器等。

2. Incremental and absolute measuring

For incremental measuring, every the unit length of motion gives a measuring signal. It has simple detecting device. Any point can be the start of the measuring. Incremental measuring is applied on most contour controlled CNC machine tools. However, if a mistake happens, no measuring could be correct afterwards.

Absolute measuring determines each displacement based on a common origin. This method is reliable. However, the complexity rises with sensitivity.

按测量的计数方式，位置检测装置可分为增量式测量和绝对式测量。

增量式测量每移动一个测量单位就发出一个测量信号。其优点是检测装置简单，任何一个对中点都可以作为测量起点，轮廓控制数控机床上大都采用这种测量方式。其缺点是如果某一步计数有误，后面的测量结果将全错。

绝对式测量可以避免上述缺点，对于被测量的任意一点位置均由固定的零点作基准，每一被测点都有一个相应的测量值。在这种测量方式下，对分辨率的要求越高，位置检测装置的结构越复杂。

3. Direct and indirect measuring

The Measuring of displacements by gratings or inducing synchronizers that are fixed on moving components is known as direct measuring. It has inherent high accuracy. Direct measuring is only applied on small machine tools since the length of the measuring device must be the same as travel distance.

Indirect measuring converts revolutions of a rotary to linear displacement of the relevant component. It has compact configuration. The measuring is not restricted by linear travel distance. However, the results are obtained through conversion, measuring accuracy therefore is limited by transfer error.

按检测装置的安装位置，位置检测装置可分为直接测量和间接测量。

直接测量是将检测装置直接安装在执行部件上。测量直线位移量常用光栅、感应同步器等检测装置。其优点是直接反映工作台的直线位移量，测量精度高；缺点是检测装置要和行程等长，这对大型数控机床是一个很大的限制。

间接测量是通过测量与工作台直线运动相关联的回转运动，间接地测量工作台的直线位移，使用可靠方便，无长度限制。其缺点是由于测量信号加入了直线运动转变为回转运动的传动链误差，从而影响测量精度。

3.4.3 Position sensors and their working principles

1. Linear gratings

Grating is also known as grating ruler (see Figure 3-48). It is a kind of accurate linear displacement sensor that is used for direct measuring of the worktable position. With laser technology and electronic technology, grating sensitivity has reached 0.1μm. Grating is a component of a closed loop control system.

光栅也称为光栅尺（图 3-48），它是一种高精度的直线位移传感器，在数控机床上用于测量工作台的位移。目前，通过激光技术制作的光栅精度达到了微米级，通过细分电路可以做到 0.1μm 甚至更高的分辨率。光栅是闭环控制系统的组件之一。

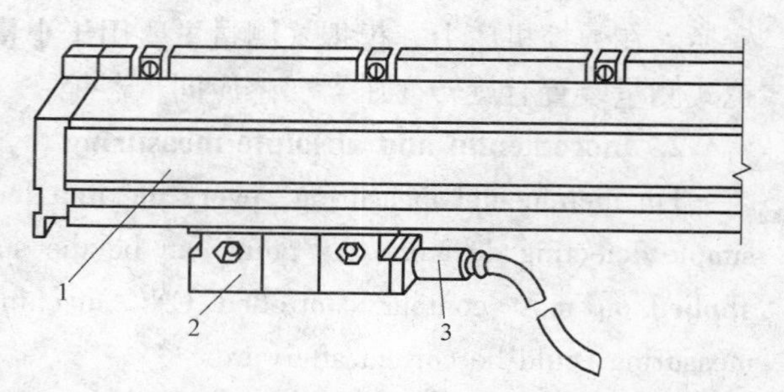

Figure 3-48 Appearance of a linear grating
1—Grating 2—Scanner 3—Cable

Linear gratings are used to measure linear displacements. Light transmits through grating can be in transmission (see Figure 3-49a) and reflection (see Figure 3-49b). Gratings are used in pairs, in which the moving component is known as the gauge grating, and the static component is the indicator grating.

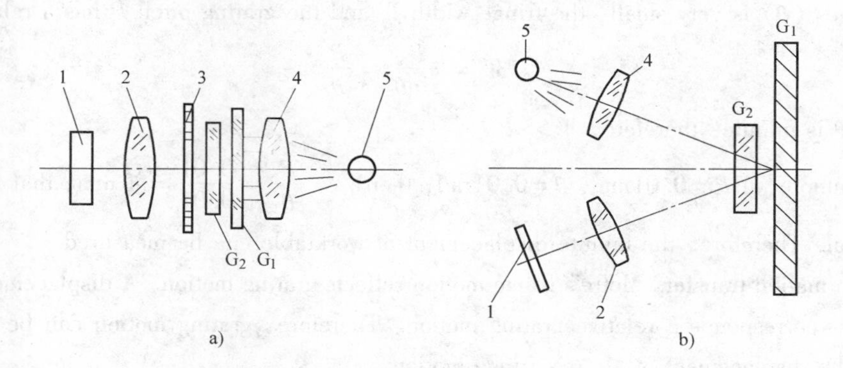

Figure 3-49 Transmission grating and reflection grating

a) Transmission grating b) Reflection grating

1—Photosensor 2、4—Lens 3—Slot 5—lamp-house

G_1—Gauge grating G_2—Indicator grating

(1) Constructions The grating is a piece of optical glass or fine polished metal board with stripes carved on. The stripes are vertical to motion direction of the grating. Usually, a grating has 50, 100 or 200 stripes per millimeter length. The distance between adjacent stripes is defined as grating pitch. Gratings are used in pairs. The moving one is known as the gauge grating, and the static one is the indicator grating.

The integral unit that consists of a lamp-house, an indicator grating and a photosensor is known as displacement-photoelectricity converter. The relative displacement between the converter and the indicator grating that is mounted on a moving component of the machine tool is converted into Moire's fringes. Photosensor reads the periodic variation of Moire's fringes to determine the quantity and direction of the displacement.

光栅是在一块长方形的光学玻璃或金属镜面上均匀地刻有许多与运动方向垂直的线纹，常用的光栅每毫米刻有 50、100 或 200 线纹。相邻线纹之间的距离称为栅距。

实际应用中，常把光源、指示光栅和光敏元件等视为一个整体，将其称为位移-光电变换器。它与标尺光栅配合产生莫尔条纹，光敏元件通过测量莫尔条纹的变化给出位移的大小和方向。

(2) Working principle The fringes on the indicator grating have a minute angle to those on the gauge grating thus producing crossings of fringes. Zones near fringe crossings have larger transparent areas thus bright. The further from the crossings the areas are darker. The alternating pattern of bright and dark area is known as Moire's fringe (see Figure 3-50).

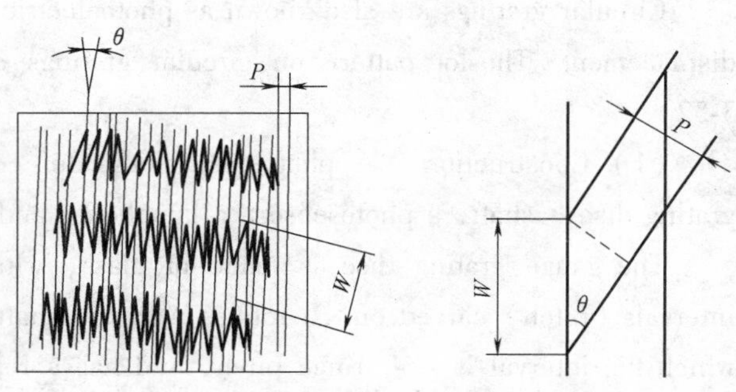

Figure 3-50 Moire's fringe

For a linear position sensor, Moire's fringe has three functions:

① Displacement magnification. When the angle between indicator grating fringes and gauge grating fringes (θ) is very small, the fringe width W and the grating pitch P has a relationship of:

$$W=\frac{P}{\sin\theta}\approx\frac{P}{\theta}$$

Since θ is minute, therefore, $W>>P$.

For example, if $P=0.01\text{mm}$, $\theta=0.01\text{rad}$, then $W\approx\frac{P}{\theta}=\frac{0.01}{0.01}=1$ mm, that is a 100-time magnification. Therefore, the minute displacement of worktable can be measured.

② Information transfer. Moire's fringe motion reflects grating motion. A displacement of Moire's fringe always corresponds a relative grating motion. Therefore, grating motion can be measured by measuring the displacement of Moire's fringe motion.

③ Error reduction. Moire's fringes are consists of thousands of brightness-alternating zones, few grating pitch error will not affect measurement.

安装时，使指示光栅上的线纹与标尺光栅上的线纹成一个很小的角度 θ。两光栅尺上线纹交叉点附近的区域内黑线重叠，透明区域大，使该区域出现亮带；而距交叉点越远的区域，两光栅不透明黑线的重叠部分越少，挡光效应增强，出现暗带。这种明暗相间且与光栅线纹几乎垂直的条纹称为莫尔条纹。

莫尔条纹具有以下特点：

① 放大作用。当两光栅尺线纹之间的夹角 θ 很小时，莫尔条纹的节距是光栅栅距的 $1/\theta$ 倍。因此莫尔条纹具有放大作用，便于测量。

② 信息变换作用。当光栅左右移动一个栅距 P 时，莫尔条纹也相应地上下准确移动一个节距 W。只要测量出莫尔条纹移过的距离，就可以得出光栅移动的微小距离。

③ 平均效应。莫尔条纹由许多明暗相间的条纹组成，对由于栅距刻制不均匀而造成的误差不敏感，提高了测量系统的可靠性。

(3) Applications

① Displacement measuring.

② Direction identifying. The order of receiving optical signals P_A and P_B represents motion direction of the worktable (see Figure 3-51).

2. Circular gratings

Circular gratings are also known as photoelectric encoders, which are used to measure angular displacement. Theslot pattern on circular gratings can be in radial and tangential (see Figure 3-52).

(1) Construction A photoelectric encoder consists of a gauge grating disc, an indicator grating disc, ashaft, a photosensor, a lamphouse and a mounting flange (see Figure 3-53).

The gauge grating disc is made of glass, with circumferential fringes that have the same intervals (pitch) carved on. It rotates with the shaft. The indicator grating has two slots between which the interval is 1/4 fringe pitch, and has a reference slot of "zero" position. The indicator grating is mounted parallel to the gauge grating disc.

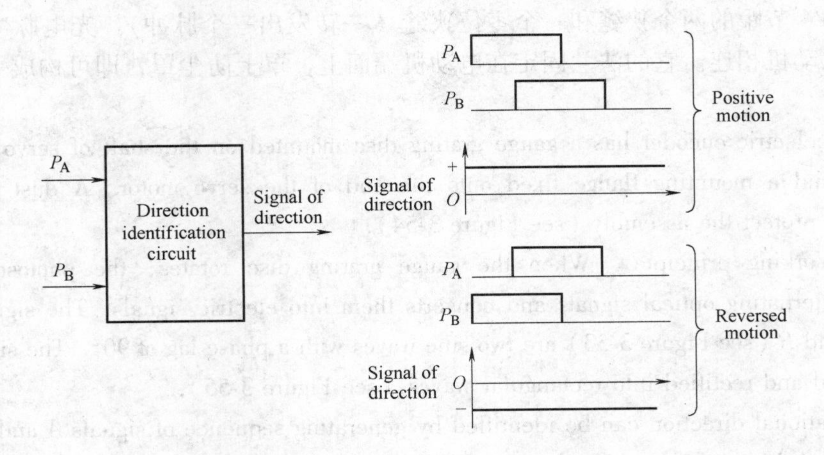

Figure 3-51 Motion direction identification

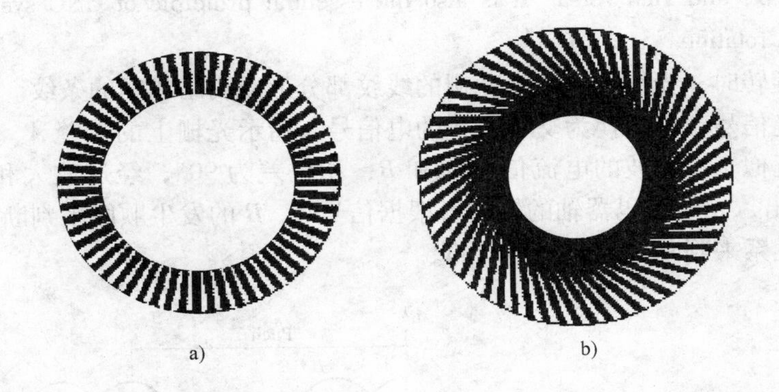

Figure 3-52 Slot patterns of gratings

a) Radial grating b) Tangential grating

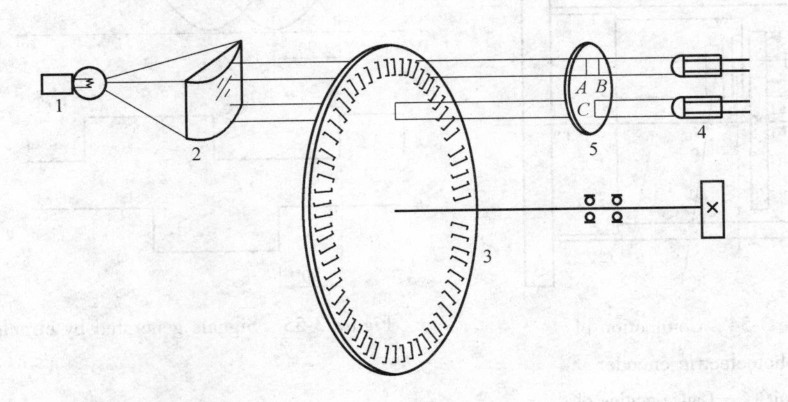

Figure 3-53 Main components of a photoelectric encoder

1—Lamp-house 2—Lens 3—Slot disc 4—Photo sensor 5—Shade

光电脉冲编码器由电路板、圆光栅、指示光栅、轴、光敏元件、光源和连接法兰等组成。圆光栅是在一个圆盘的圆周上刻有相等间距的线纹，分为透明部分和不透明部分。圆光栅与工作轴一起旋转。与圆光栅相对平行地放置一个固定的扇形薄片，称为指示光栅，上面刻有相差 1/4 节距的两个狭缝和一个零位狭缝（一转发出一个脉冲）。光电脉冲编码器通过键与伺服电动机相连。它的法兰固定在电动机端面上，罩上防尘罩，即可构成一个完整的检测装置。

A photoelectric encoder has a gauge grating disc mounted on the shaft of servo motor by key connection and a mounting flange fixed onto the end of the servo motor. A dust-proof cover is necessary to protect the assembly (see Figure 3-54).

(2) Working principle When the gauge grating disc rotates, the photosensor receives brightness-alternating optical signals and converts them into electric signals. The signals generated by slots *A* and *B* (see Figure 3-53) are two sine waves with a phase lag of 90°. The sine waves then are magnified and rectified into rectangular waves (see Figure 3-55).

The rotational direction can be identified by generating sequence of signals *A* and *B*. In Figure 3-53, signal *A* is received earier than signal *B* by 90°, that means the shaft is rotating in the direction of *A* to *B*, and vice versa. It is also the essential principle of CNC system identifing the direction of shaft rotation.

当圆光栅旋转时，光线透过两个光栅的线纹部分形成明暗相间的条纹。光敏元件接收这些明暗相间的光信号，并转换为交替变化的电信号。指示光栅上的狭缝 *A*、*B*（图 3-53）分别生成为两组近似于正弦波的电流信号 *A* 和 *B*，相位差为 90°，经过放大和整形变成方波。由图 3-55 可看出，光电编码器轴的转向可根据信号 *A*、*B* 的发生顺序来判断。数控系统正是利用这一相位关系来判断轴的旋转方向的。

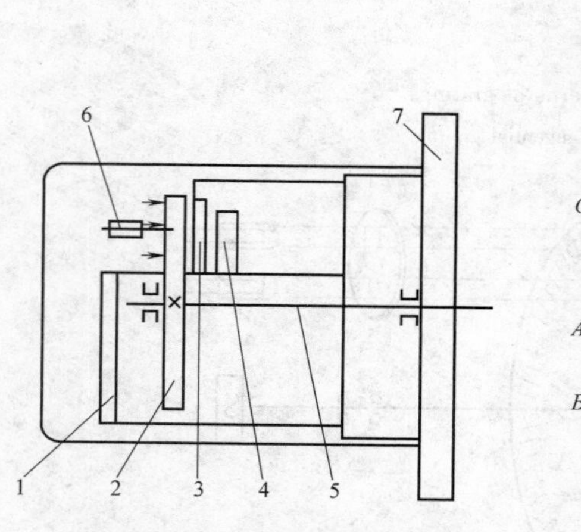

Figure 3-54 Connection of a photoelectric encoder

1—Circuit 2—Gauge grating disc
3—Indicator grating disc 4—Photosensor
5—Shaft 6—Lamphouse 7—Mounting flange

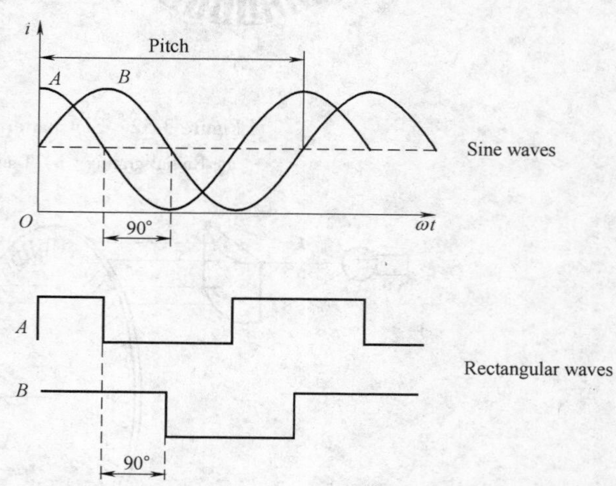

Figure 3-55 Signals generated by circular gratings

(3) Applications on CNC machine tools

① Displacement measuring. Photoelectric encoder measures the displacement of moving components by counting the brightness alternations. The counting results will be sent to CNC system

for comparing and correction.

② Tachometer. Rotational speed is calculated by frequency of optical pulse during a period, i. e.

$$\text{rpm} = \frac{N_1}{N} \times \frac{60}{t}$$

Where:

t is the sample period (in second); N_1 is the number of pulses generated during period t; N is the number of pulses in each revolution of the encoder.

(4) Accuracy improvements (see Figure 3-56) If a ball screw of 8mm pitch is driven by a motor that equipped with an encoder of 2000 pulses/rev, then the accuracy will be:

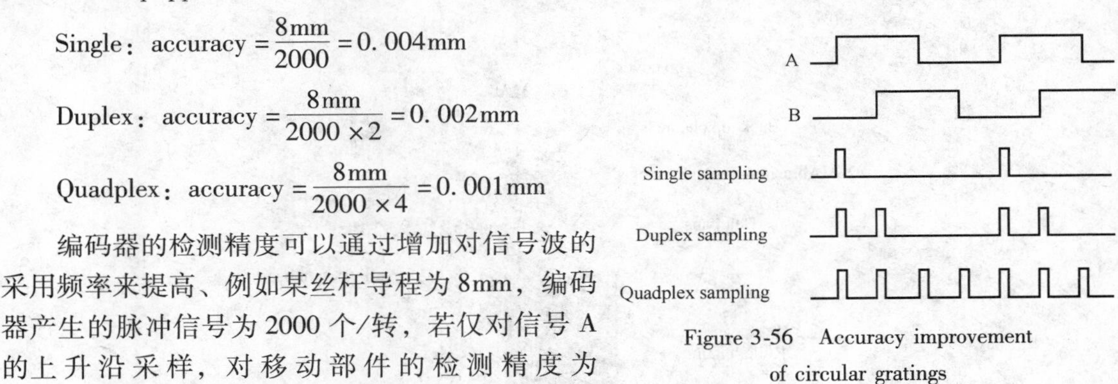

Single: accuracy $= \frac{8\text{mm}}{2000} = 0.004\text{mm}$

Duplex: accuracy $= \frac{8\text{mm}}{2000 \times 2} = 0.002\text{mm}$

Quadplex: accuracy $= \frac{8\text{mm}}{2000 \times 4} = 0.001\text{mm}$

Figure 3-56 Accuracy improvement of circular gratings

编码器的检测精度可以通过增加对信号波的采用频率来提高、例如某丝杆导程为8mm，编码器产生的脉冲信号为2000个/转，若仅对信号A的上升沿采样，对移动部件的检测精度为0.004mm；若对信号A的上升沿、下降沿分别采样，检测精度提高至0.002mm；若对信号A、B的上升沿、下降沿均进行采样，检测精度进一步提高至0.001mm。

Photoelectric encoders have no physical contact between components during operation thus eliminating friction, abrasion and noise to ensure a long service life. The system is smart and compact, therefore, it requires low driving torque to save energy. Light transmission is ultra fast, which gives the system rapid response. However, optical components are fragile and likely to be contaminated.

光电编码器工作时，元件之间没有机械接触，因此无摩擦、无磨损、无噪音，使用寿命长，响应快，同时消耗的能量极少。但使用时要防止撞击。

3.5 Automatic chip remover

3.5.1 Necessities of automatic chip removers

CNC machine tools have high machining efficiency, as a result, the quantity of material which has been cut off increases as well. If the accumulated chips are not effectively cleaned, they may twist on the workpiece to affect automatic machining. Moreover, thermal distortion of the machine tool or the workpiece due to chips' heat radiation affects machining accuracy. Therefore, chip removing during machining is important for CNC machine tools. The task of chip remover involves transferring the chips out of the machine tool, and draining the cutting fluid from the chips to the reservoir.

数控机床加工效率高，单位时间内金属切削量高于普通机床，使切屑所占的空间也成倍增大。这些切屑如果不及时清除，会覆盖或缠绕在工件上阻碍自动加工。此外，热的切屑还会向机床或工件散发热量，使机床或工件产生变形而影响加工精度。因此必须迅速有效地排除切屑。排屑装置的主要作用是将切屑从加工区域排出到数控机床之外，并将切削液从切屑中分离出来，回收到切削液箱。

3.5.2 Typical automatic chip removers

Figure 3-57 shows a typical chain chip remover.

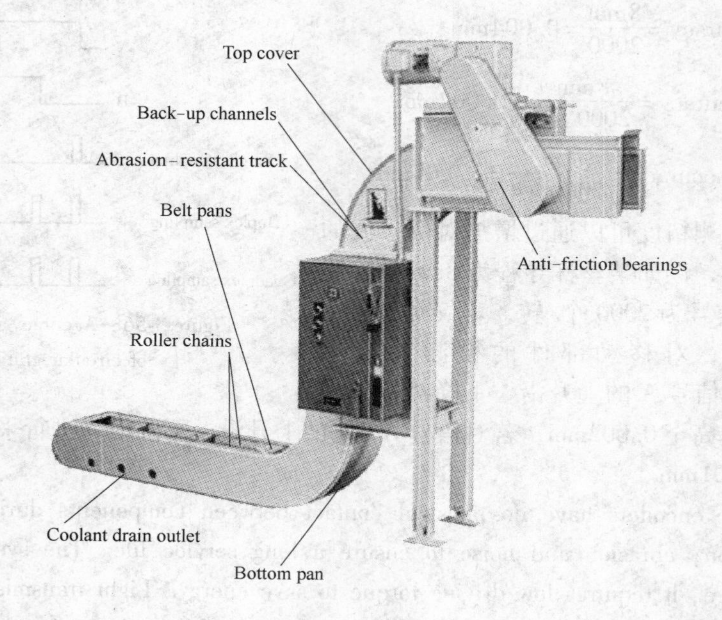

Figure 3-57 A typical chain chip remover

1. Chain chip remover（平板链式自动排屑装置）

Chain chip removers (see Figure 3-58) are widely used because they are compatible with all kinds of machine tools, and can remove any types of chips.

2. Scraper chip remover（刮板式自动排屑装置）

Scraper chip removers (see Figure 3-59) have similar construction and work principle to those of chain chip removers. The scrapers on the chain are especially suitable for removing small size chips.

3. Screw conveyor chip remover（螺旋式自动排屑装置）

The rotation of the screw drives the chip to move axially. The screw blade is welded onto the shaft. Screw conveyor chip removers (see Figure 3-60) have compact construction so that require less space. However, vertical elevation and corner transfer are impossible.

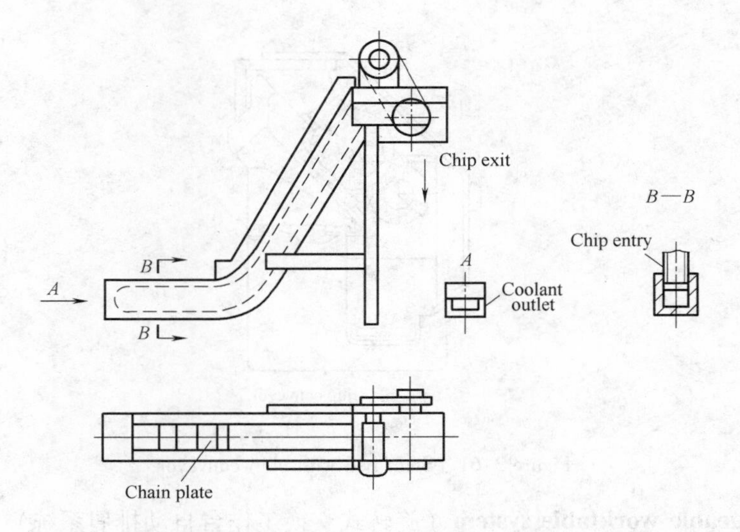

Figure 3-58 Construction of chain chip remover

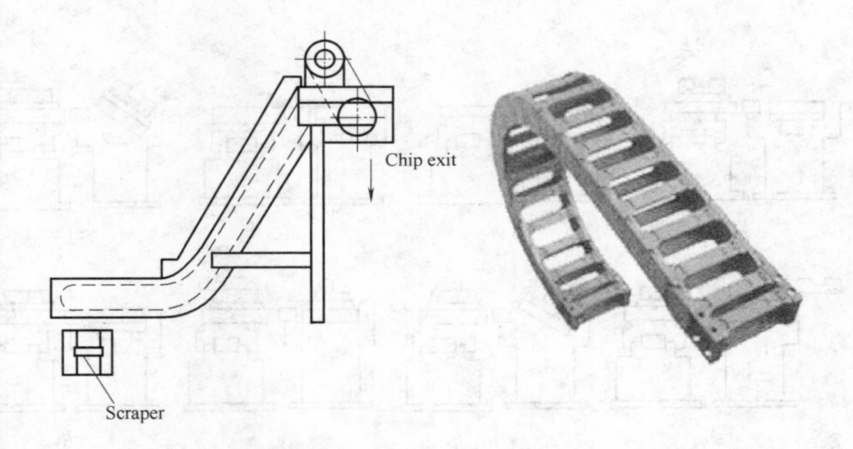

Figure 3-59 Scraper chip remover

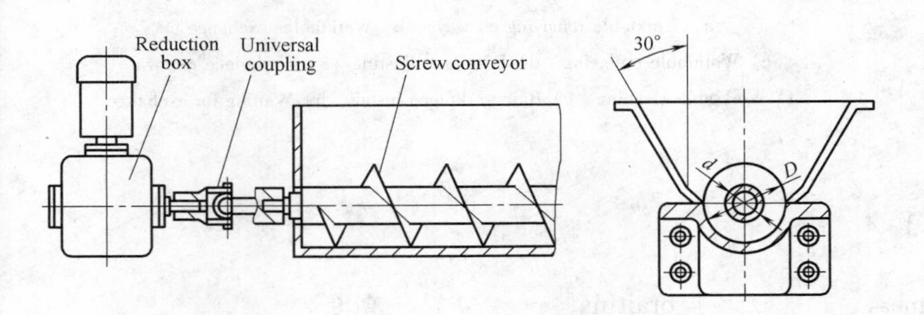

Figure 3-60 Screw conveyor chip remover

4. Slant bed with chip conveyor（倾斜式床身及切屑传送带排屑装置）（see Figure 3-61）

Chips slide down along theslant bed onto conveyor due to gravity. It provides the simplest mean of chip removing but only is applied on medium or small size lathes.

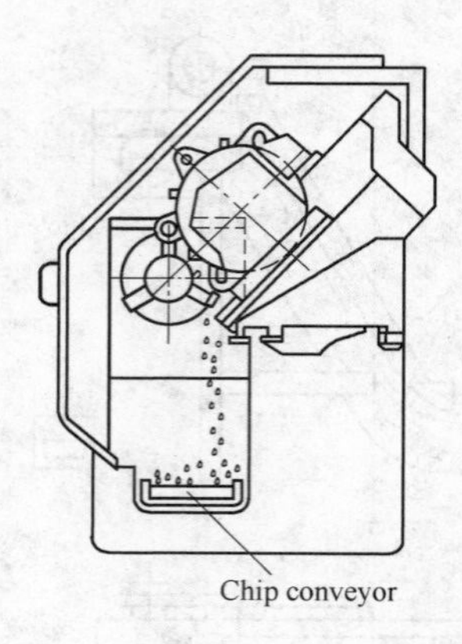

Figure 3-61　Slant bed with chip conveyor

5. Exchangeable worktable system（旋转式交换工作台自动排屑系统）

Figure 3-62 shows the working procedure of an exchangeable worktable system.

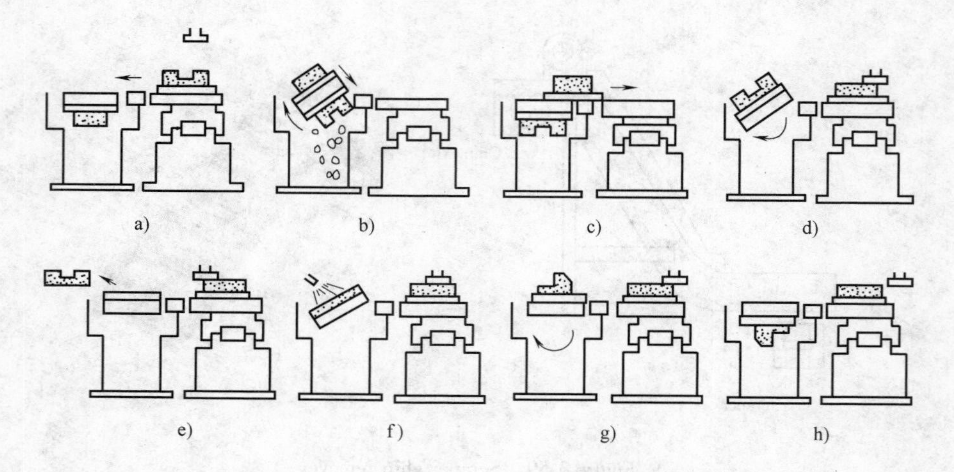

Figure 3-62　Exchangeable worktable system

a) Worktable removing off ways　b) Worktables exchange

c) Worktable replacing　d) Position adjusting　e) Workpiece removal

f) Worktable cleaning　g) Raw workpiece install　h) Waiting for exchange

Glossary

brightness	[ˈbraitnis]	亮度
chip remover		排屑装置
clearance	[ˈkliərəns]	间隙
component	[kəmˈpəunənt]	（机器、设备等的）零件；部件
conical	[ˈkɔnikəl]	圆锥形的

designate	[ˈdezigneit]	委派；指定
dovetail	[ˈdʌvˌteil]	楔形榫头；燕尾槽
eccentric	[ikˈsentrik]	不同圆心的
engage	[inˈgeidʒ]	使（齿轮等）啮合
exchangeable	[iksˈtʃeindʒəbl]	可替换的；可交换的
grating	[ˈgreitiŋ]	格栅；光栅
helical groove		螺旋状沟槽
imbalance	[imˈbæləns]	不均衡状态
incremental	[ˌinkriˈməntl]	增量的；增加的
inertia	[inˈəːʃiə]	惯性；惯量
inherent frequency		固有频率
innovative	[ˈinəuveitiv]	创新的
linear	[ˈliniə]	直线的；线性的
lubrication	[ˌluːbriˈkeiʃən]	润滑
Moire's fringe		莫尔条纹
moment of inertia		转动惯量
optimum	[ˈɔptiməm]	优化的；最佳的
orientation	[ˌɔːrienˈteiʃən]	定位；定向
preload	[priˈləud]	预紧力
quenching	[kwentʃiŋ]	淬火；急冷
rack and pinion		齿条齿轮（传动装置）
reflection	[riˈfiekʃən]	反射；映像
respondence	[risˈpɔndəns]	反应；响应
rpm		每分钟转数（revolution（s）per minute）
shim	[ʃim]	薄垫片；填隙片
staggered	[ˈstægəd]	交错的；参差的
tachometer	[tæˈkɔmitə]	转速计
Teflon	[ˈteflɔn]	聚四氟乙烯，特氟隆（商标名）
transmission	[trænsˈmiʃən]	传播；传动
versatile	[ˈvəːsətail]	多功能的；多才多艺的
vibration	[vaiˈbreiʃən]	颤动；振动

Exercises

1. List the functions of spindle orientation, and explain the work principle of the system shown in Figure 3-11.

2. In a differential teeth clearance adjusting system, one of the gears has 60 teeth, and the other has 61. Determine the minimum displacement that can be precisely controlled between the nuts.

3. If a pair of gratings have 100 fringes/mm, and the angle between the fringes of the gratings $\theta = 0.01\mathrm{rad}$, determine the fringe width W.

4. Explain the function of components 4 and 5 in Figure 3-24.

5. Chips may accumulate without chip remover. What are the negative effects?

Chapter 4　CNC Lathe

4.1 Introduction of CNC lathes

4.1.1 Capabilities of CNC lathes

Similar to conventional lathes, CNC lathes are used to machine revolutionary surfaces on shafts, bushes, plates, e.g. inner or outer cylindrical surfaces, conical surfaces, threads. However, CNC lathes have better machining accuracy (dimension accuracy up to IT5 or IT6, and surface roughness $Ra \leqslant 1.6\mu m$), and are capable to machine complex contours.

Turning cutters, drills, reamers, boring cutters and thread cutters are commonly used cutting tools for CNC lathes.

数控车床主要用于加工各种轴类、套筒类和盘类零件的回转表面，如内外圆柱面、圆锥面、成形回转表面、螺纹面以及其他高精度的曲面与端面螺纹。数控车床加工的公差等级可达 IT5 或 IT6，表面粗糙度值 Ra 为 1.6μm 以下。

4.1.2 Classifications of CNC lathes

With the development of machine tool manufacturing, CNC lathes can be classified by their capabilities.

(1) Economy CNC lathes　The design of economy CNC lathes (see Figure 4-1) is usually a modification based on that of conventional lathes. They are equipped with open loop servomechanisms, and use single board microcomputers as their CNC devices. Economy CNC lathes

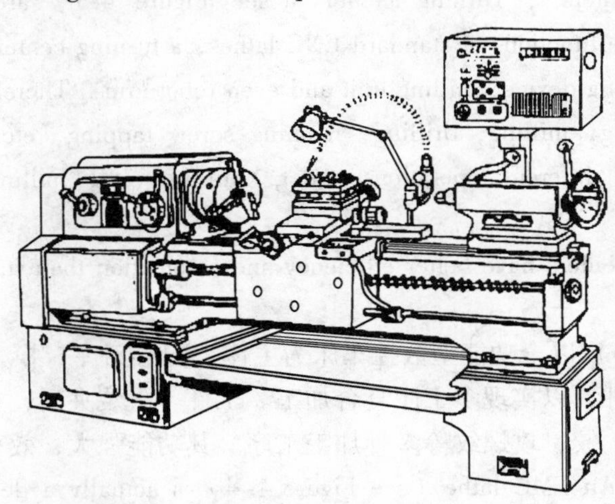

Figure 4-1　An economy CNC lathe

have simple constructions and are less expensive. Compared to other CNC lathes, they lack of the functions of automatic tool tip radius compensation and constant linear velocity control.

经济型数控车床（图4-1）一般是在普通车床的基础上进行改进设计，并采用步进电动机驱动的开环伺服系统，其控制部分采用单板机或单片机。此类车床结构简单，价格低廉，但无刀尖圆弧半径自动补偿和恒线速度切削等功能。

（2）Full-function CNC lathes　Full-function CNC lathes are known as CNC lathes or standard CNC lathes (see Figure 4-2), representing standard control system. It has a high resolution CRT. Graphics simulation, tool compensation, communication/network and multi-axis controllability are the standard functions of the system. Full-function CNC lathes have closed loop or semi-closed loop controllers for accuracy. Besides, high rigidity and high efficiency are highlighted.

全功能型数控车床通常简称数控车床，又称为标准型数控车床，如图4-2所示。其数控系统是标准型的，带有高分辨率的CRT显示器，具备各种显示、图形仿真、刀具补偿等功能及通信或网络接口。全功能型数控车床采用闭环或半闭环控制的伺服系统，可以进行多轴控制，具有高刚度、高精度和高效率等特点。

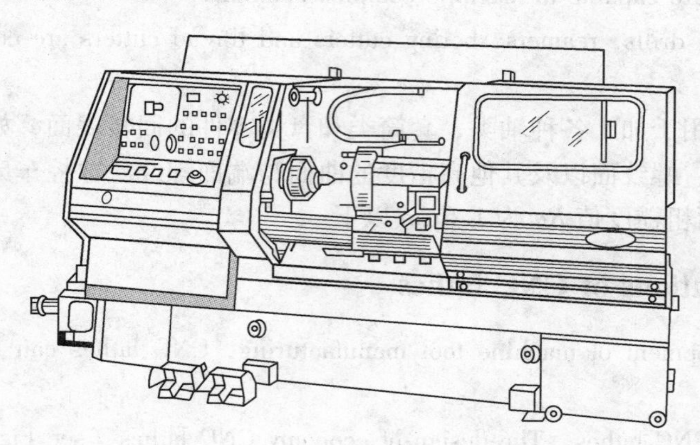

Figure 4-2　A standard CNC lathe

（3）Turning centers　Turning centers (see Figure 4-3) are designed to increase productivity. Besides the functions of standard CNC lathes, a turning center is equipped with a tool magazine, ATC, indexing device, milling unit and even robot arms. Therefore, different machining process such as turning, milling, drilling, reaming screw tapping, etc. can be performed by clamping once. Even different operations (e.g. turning and milling) can be performed simultaneously.

Although turning centers have better efficiency and automation than standard CNC lathes, they are costly.

车削中心（图4-3）以全功能型数控车床为主体，配有刀库、换刀装置、分度装置、铣削装置和机械手等部件，以实现多工序复合加工。工件一次装夹后，在车削中心可完成回转类零件的车、铣、钻、铰、攻螺纹等多种加工工序。其功能强大，效率高，但价格较贵。

（4）FMC lathe　An FMC lathe (see Figure 4-4) is actually a flexible manufacturing cell consists of a CNC lathe and a robot. It is capable to perform automatic transfer, clamping and

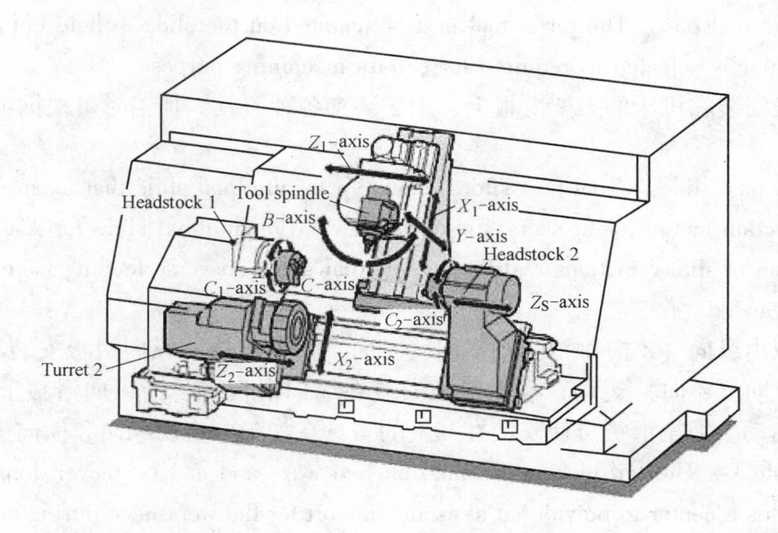

Figure 4-3 A multi-axis turning center

machining of a workpiece. Preparations and adjusting are also automatic.

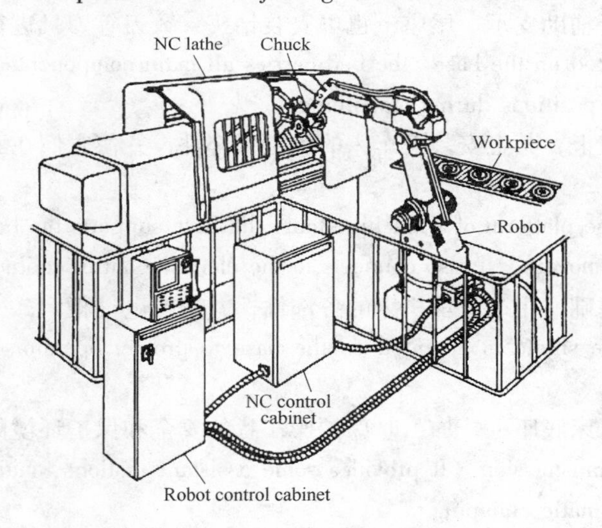

Figure 4-4 An FMC lathe

FMC 车床（图4-4）是一个由数控车床、机器人等构成的柔性加工单元。它能实现工件搬运、装卸的自动化和加工调整准备的自动化。

4. 1. 3 Construction and features of CNC lathes

1. Construction of CNC lathes

（1）Headstock The headstock locates on the left of the bed. The main function of the headstock is supporting the spindle and accommodating spindle transmissions for the primary motion of the lathe.

主轴箱固定于床身的最左边，用于支承并带动主轴，以实现机床的主运动。

(2) Turret tool post The turret tool post is mounted on the slide to hold cutting tools, which can be automatically selected as required for certain machining purpose.

转塔刀架安装在机床的刀架滑板上，用于装夹各种刀具。加工时可根据加工要求自动换刀。

(3) Tool post slides Tool post slides include a longitudinal slide that locates on bed ways to perform Z-direction motion, and a cross slide locates on longitudinal slide for X-direction motion. The combination of slides' motions realizes longitudinal and transverse feeding motion of the cutting tools that are carried.

刀架滑板由纵向（Z 向）滑板和横向（X 向）滑板组成。纵向滑板安装在床身导轨上，沿床身实现纵向（Z 向）运动；横向滑板安装在纵向滑板上，沿纵向滑板上的导轨实现横向（X 向）运动。刀架滑板可使安装在其上的刀具在加工中实现纵向和横向进给运动。

(4) Tailstock The tailstock is mounted on bed ways and can be moved longitudinally along the ways. It holds a center to provide an assistant support for the workpiece during machining. It can also carry tools (e. g. drills, reamers) for hole making.

尾座安装在床身导轨上，可沿导轨进行纵向移动以调整位置。尾座主要用于安装顶尖，在加工中对工件进行辅助支承。尾座上也可安装钻头、铰刀等刀具进行孔加工。

(5) Bed Mounted on the base, the bed carries all main components of the lathe and ensures their accurate relative positions during machining.

床身固定在底座上，其上安装着车床的各主要部件，并使它们在工作时保持准确的相对位置。

(6) Base As the platform of a machine tool, the base supports the bed and other components (e. g. shield, chip remover). It also connects to the electric control cabinet.

底座是车床的基础，用于支承机床的各部件，并连接电气柜。

(7) Shield The shield is mounted on the base to protect operators and the workshop from chip.

防护罩安装在机床底座上，加工时保护操作者的安全和保护环境的清洁。

(8) Hydraulic transmission It provides some assistant motions, mainly, spindle gear shift, tailstock moving, automatic clamping.

机床液压传动系统用来实现机床上的一些辅助运动，主要是实现机床主轴的变速、尾座套筒的移动及工件自动夹紧机构的动作。

(9) Lubricating system The lubricating system provides both lubrication and cooling for mechanical components of the machine tool.

机床润滑系统为机床运动部件提供润滑和冷却。

(10) Cutting fluid supply It circulates cutting fluid for machining, i. e. pumps and filters cutting fluid to the location where cutting is performing, and recycles the fluid after the drainage.

机床切削液供给系统为切削液循环提供动力，为机床在加工中提供充足的切削液，以满足切削加工的要求。

(11) Electric control system The electric control system consists of a CNC system (including a CNC device, a servo system and a PLC) and an electrical control system. It realizes automatic

machine tool control.

机床的电气控制系统主要由数控系统（包括数控装置、伺服系统及可编程序控制器）和机床的强电气控制系统组成，它能完成对机床的自动控制。

2. Characteristics of modern CNC lathes

Compare to conventional lathes, CNC lathes have the advantages of:

① High accuracy. Due to improved performance of control systems and optimized mechanical construction.

② High efficiency. Due to innovative tool materials and optimized mechanical components (e. g. rigidity, spindle rpm, rated power). It is 2-3 times efficient than conventional horizontal lathes. With the increasing complexity of the workpiece, its efficiency becomes more obvious.

③ High flexibility. More than 70% of all kinds of workpieces (in small batch production) can be automatically machined.

④ High reliability. Time of operating without failure reaches up to 30, 000 hours.

⑤ Versatility. Both rough and finish machining can be performed. All or most machining processes can be completed by clamping once.

⑥ Modularized design (see Figure 4-5). A lathe consists of functional modules that can be replaced as required to perform different machining tasks.

与普通车床相比，数控车床具有高精度、高效率、高柔性、高可靠性、工艺能力强及模块化结构等特点。

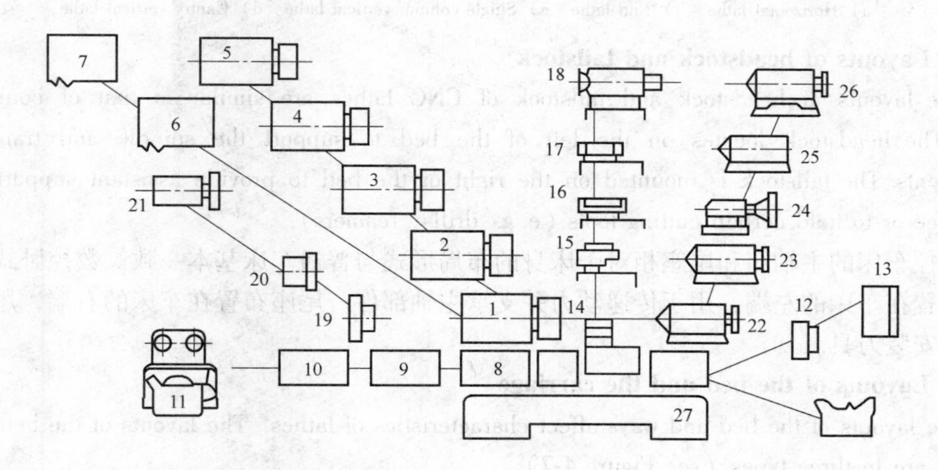

Figure 4-5 Modularized components of a CNC lathe

1 ~ 7—Headstock modules 8 ~ 11—Feeding mechanism

12 ~ 13—Fast moving mechanisms 14 ~ 18—Tool post modules

19 ~ 21—Chuck modules 22 ~ 26—Tailstock modules 27—Bed

4. 1. 4 Layout of CNC lathes

1. Factors that determine layouts

The layout of lathes mainly depends on dimensions, quality requirements and profiles of workpieces that need to be machined. Manufacturing efficiency, accuracy and safety should be considered as well. Figure 4-6 shows some typical layouts of CNC lathes.

数控车床的布局形式受工件尺寸、质量和形状，以及机床生产率、机床精度、安全要求等因素的影响。根据布局形式，常见的数控车床有卧式车床、卡盘车床、单/双柱立式车床和龙门移动式立式车床等形式，如图 4-6 所示。

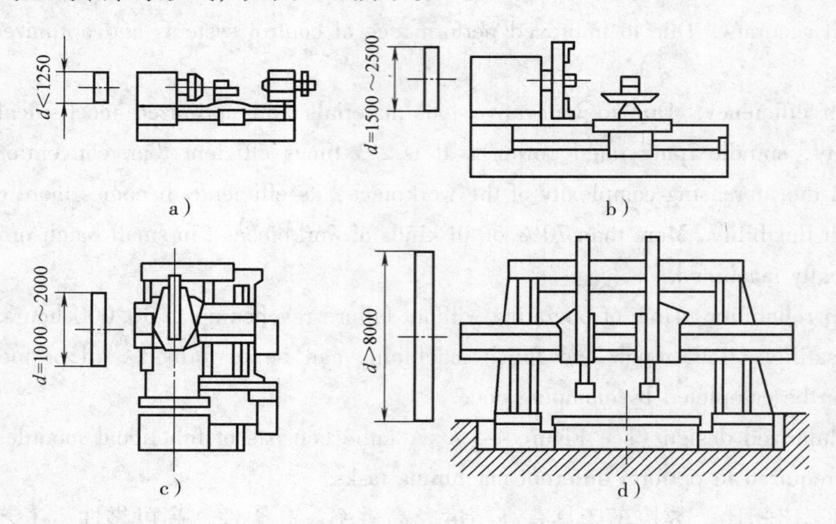

Figure 4-6 Relationship between layout and workpiece dimensions

a) Horizontal lathe b) End lathe c) Single column vertical lathe d) Gantry vertical lathe

2. Layouts of headstock and tailstock

The layouts of headstock and tailstock of CNC lathes are similar to that of conventional lathes. The headstock locates on the left of the bed to support the spindle and transmission components. The tailstock is mounted on the right of the bed to provide assistant support for the workpiece or to hold certain cutting tools (e. g. drills, reamers).

数控车床的主轴箱和尾座相对于床身的布局形式与普通车床基本一致。数控卧式车床主轴箱布置在车床的左端，用于传递动力并支承主轴部件；尾座布置在车床的右端，用于支承工件或安装刀具。

3. Layouts of the bed and the carriage

The layouts of the bed and ways affect characteristics of lathes. The layouts of the bed and the carriage are in three types (see Figure 4-7).

(1) Horizontal bed-horizontal carriage (Figure 4-7a) The horizontal bed has good manufacturability for ways. The horizontal tool post mounted on the horizontal bed provides better motion accuracy. This type of layout is widely applied on large CNC lathes or small precise CNC lathes.

However, insufficient space under the bed affects chip removing. Moreover, the horizontal carriage increases lateral dimension of the lathe.

水平床身工艺性好，便于导轨面的加工。水平床身配水平放置的刀架可提高刀架的运动精度，一般用于大型数控车床或小型精密数控车床的布局。但是水平床身下部空间小，影响

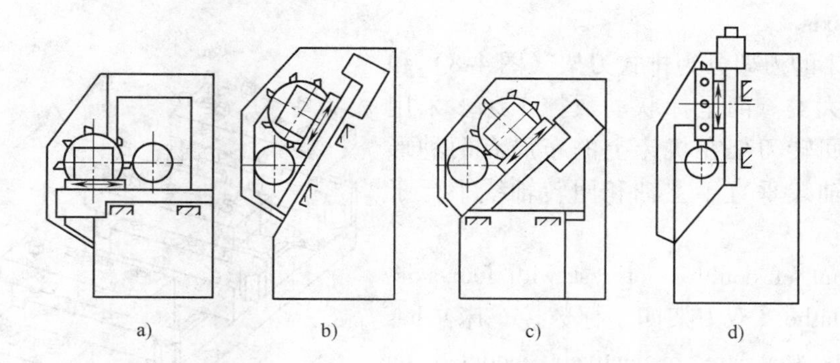

Figure 4-7 Layouts of CNC horizontal lathes
a）Horizontal bed-horizontal carriage b）Slant bed-slant carriage
c）Horizontal bed-slant carriage d）Vertical bed-vertical carriage

排屑。水平滑板横向尺寸较大，从而加大了机床宽度方向的尺寸。

（2）Slant bed-slant carriage（see Figure 4-7b，d） The inclination of the bed can be in 30°，45°，60°，75° and 90°，where a 90° bed inclination is known as vertical bed-vertical carriage. Small inclination affects chip removing and also increases lateral dimension of the lathe，and large inclination affects guidance accuracy and force distribution on ways. The inclination also determines height-width ratio of a lathe. Therefore，medium and small size CNC lathes usually have 60°-inclination beds as a compromise.

倾斜床身-倾斜滑板结构的导轨倾斜角度有 30°、45°、60°、75°和 90°等几种，其中 90°的称为直立床身-直立滑板。若倾斜角度过小，则排屑不便；若倾斜角度过大，则导轨的导向性及受力情况差。导轨倾斜角度的大小还直接影响机床外形高宽比。综合考虑诸项因素，中小规格的数控车床宜采用倾斜角为 60°的床身。

（3）Horizontal bed-slant carriage（see Figure 4-7c） The layout has both advantages of the horizontal bed and the slant bed. Therefore，it is regarded as the optimum layout for horizontal CNC lathes. Ways are shielded to keep contaminants out.

水平床身-倾斜滑板这种布局形式具有水平床身工艺性好的特点，且排屑方便，结构紧凑，一般认为是卧式数控车床的最佳布局形式。这种结构通常配置有倾斜式的导轨防护罩。

Slant carriage can remove chip by gravity. It also has enough space under carriage to accommodate automatic chip remover. The compact construction requires less area for lathe mounting. In a FMC lathe，the slant carriage is suitable for a robot arm action. Closed-protection is convenient to be realized as well.

倾斜滑板的布局普遍用于中、小型数控车床，因为排屑容易，热切屑不会堆积在导轨上，也便于安装自动排屑装置。这种布局形式结构紧凑，机床占地面积小，便于机械手动作，有利于实现单机自动化，整个车床容易实现封闭式防护。

4. Layout of tool rest

Both linear tool rest（see Figure 4-8） and rotary tool rest are applied on CNC lathes，but the latter is commonly used on two-axis control CNC lathes. According to the relative position between the tool rest axis and the spindle axis，rotary tool rests are in two forms，namely，perpendicular axis and parallel axis.

数控车床的刀架分为排式刀架（图4-8）和回转刀架两大类。两坐标联动数控车床多采用回转刀架。回转刀架在机床上的布局有两种形式，即回转轴线垂直于主轴和回转轴线平行于主轴。

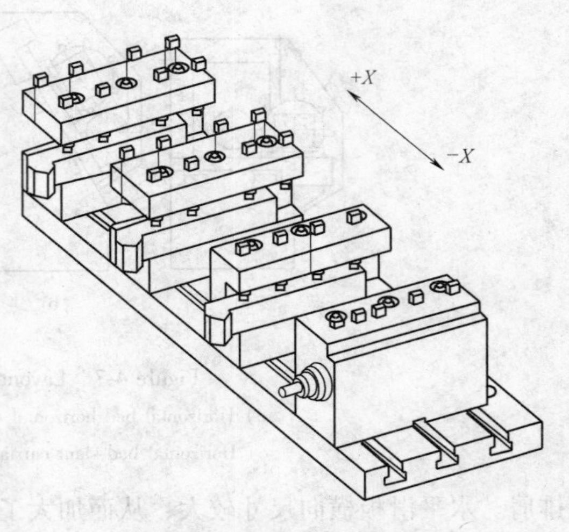

Figure 4-8 Linear tool rest

The layout of double tool rest with four-axis-control CNC lathe（双刀架四坐标数控车床）has two tool rests that are respectively mounted on different carriages（see Figure 4-9）. The tool rests are independently controlled to machine different position of a workpiece at the same time. The layout is versatile and efficient, especially suitable for machining complex parts in mass production, e. g. crank shafts, aircraft components, etc.

双刀架（图4-9）四坐标数控车床有两个独立的滑板和回转刀架，切削过程中同一时刻，刀具相对于工件的位置是独立的，每个刀架的进给量是分别控制的，因此，两刀架上的刀具可以同时切削同一工件的不同部位，既扩大了加工范围，又提高了加工效率，适合于加工曲轴、飞机零件等形状复杂、批量较大的零件。

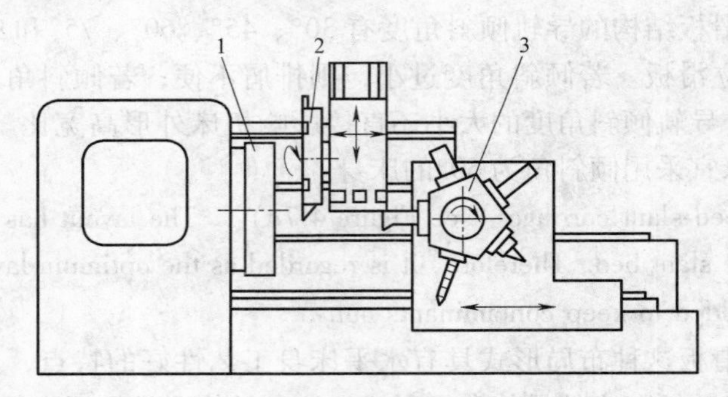

Figure 4-9 Double tool rest layout

1—Spindle 2—Upper tool post 3—Lower tool post

4.2 Transmission of CNC lathe

4.2.1 Primary transmission

1. Requirements

Spindle transmission is expected to have a wide spindle rotation speed range and continuous variable rotation speed for optimum cutting speed. Sufficient driving power is also desired to overcome cutting force. The spindle should have rigid construction and anti-vibration ability, and an accurate and stable rotary axis.

Spindle rotation speed of a CNC lathe is set by program. To reduce noise and vibration, also to improve transmission accuracy, the transmission chain is expected to be as short as possible. To improve product efficiency and quality, constant linear velocity control is an important function of a CNC lathe. Moreover, for convenience, the configuration of the spindle end with its chuck should enable automatic clamping.

数控车床的主轴传动速度可调范围要大，驱动功率要足够大，主轴回转轴线的位置应准确稳定，有较高刚性与抗振性。数控车床的主轴变速是根据加工程序指令自动进行的。为了实现降噪减振，并确保主轴传动精度，主传动链要尽可能地短。为了保证生产率和加工质量，主轴传动应能实现恒切削速度控制。主轴及卡爪还应能配合其他构件实现工件的自动装夹。

Figure 4-10 is the transmission diagram of a TND360 horizontal CNC lathe. The primary motion of the lathe is driven by a DC servomotor (rating power of 27kW). The gears provide the spindle low speed zones (7 ~ 800r/min) and high speed zones (800 ~ 3150r/min) by gear shifting between shaft Ⅰ and shaft Ⅱ (see Figure 4-11). In both speed zones, spindle rotation speed can be continuous varied by the DC motor without gears thus improved transmission accuracy.

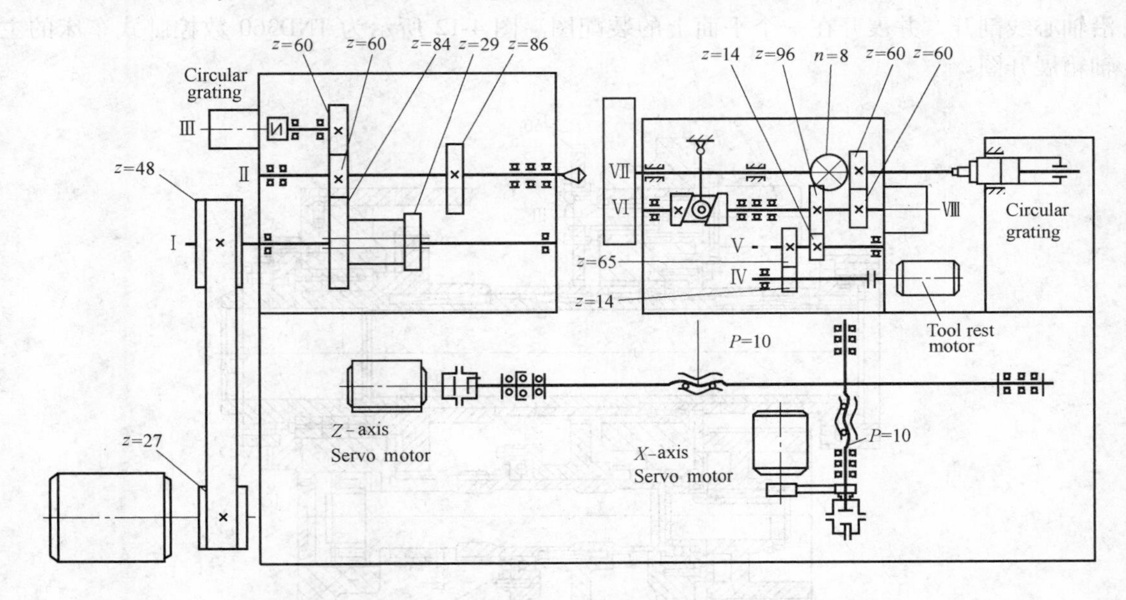

Figure 4-10 Transmission diagram of a TND360 horizontal CNC lathe

The spindle rotational speed is monitored by a circular grating, which is connected to the spindle shaft with a transmission ratio of 1∶1 through a pair of gears. The circular grating reads spindle rotation speed and sends the information as electrical signals to CNC device.

For CNC lathes, thread turning requires no mechanical connection between spindle rotation and *Z/X*-axis feeding. The pitch is precisely controlled by spindle rotation speed and corresponding feedrate of the cutting tool.

TND360 数控卧式车床的主运动由主轴直流伺服电动机（27kW）驱动。双联滑移齿轮使主轴获得高、低两挡转速范围。在各转速范围内，伺服电动机可实现主轴的无级变速。由于省去了大部分齿轮传动变速机构，因此减小了齿轮传动对主轴精度的影响。主轴运动经过传动比为 1:1 的齿轮副驱动圆光栅，圆光栅将主轴的转速转变为电信号发送给数控装置。

数控车床上螺纹切削加工的方式与普通机床是不同的。它可实现主轴每转一圈，进给轴 Z 轴或 X 轴移动一个导程。

2. Construction of spindle case

Spread chart is often used to illustrate construction and components of a spindle case. A spread chart is composed by a series of unfolded sections that through transmission shafts' axis. Figure 4-12 is the spread chart of the spindle case of the TND360.

表达机床主轴箱的结构和各传动元件装配关系时常用展开图。展开图是按传动链传递运动的先后顺序，沿轴心线剖开，并展开在一个平面上的装配图。图 4-12 所示为 TND360 数控卧式车床的主轴箱展开图。

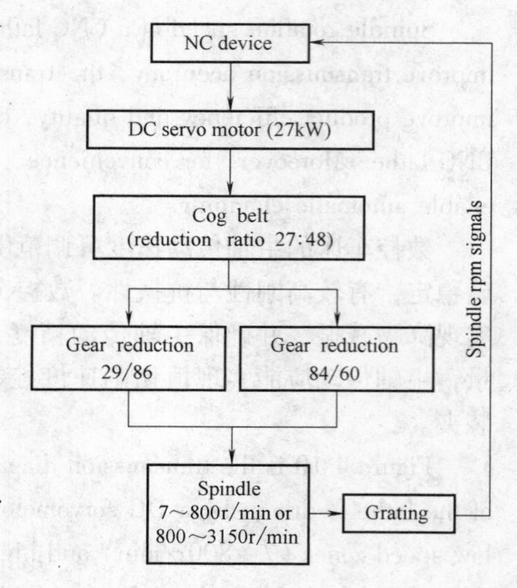

Figure 4-11 Transmission chain of the TND360

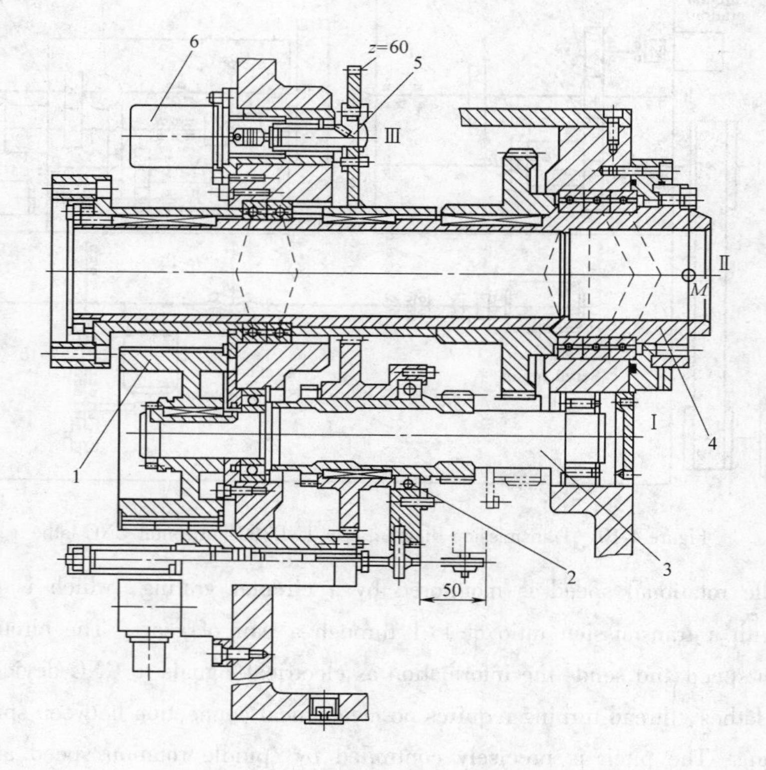

Figure 4-12 Primary transmission construction of TND360

1—Pulley 2—Fork bearing 3—Shifting shaft
4—Spindle 5—Tachometer shaft 6—Circular grating

(1) Spindle motor The spindle of the TND360 is driven by a stepless DC servo motor, which has rating speed of 2000r/min and maximum 4000r/min. A tachometer is attached at the end of the motor for precise speed control. The motor runs between 35r/min and 2000r/min with a constant torque output by voltaic adjusting, and runs between 2000r/min and maximum with a constant power output by magnetic adjusting (see Figure 4-13).

主轴采用直流伺服电动机驱动，无级调速，由安装在电动机尾部的测速器实现速度反馈。额定转速为 2000r/min，最高转速为 4000r/min，最低转速为 35r/min。额定转速至最高转速之间为调磁调速，有恒功率输出；最低转速至额定转速之间为调压调速，有恒转矩输出，如图 4-13 所示。

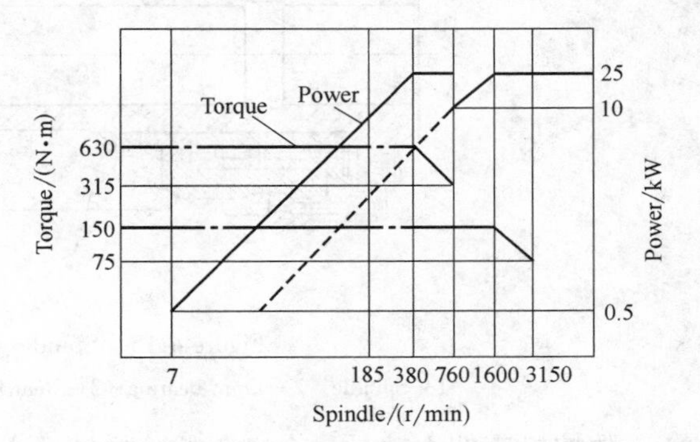

Figure 4-13 Power-torque diagram of a spindle motor

(2) Gear shifting shaft (shaft Ⅰ) As shown in Figure 4-14, the gear shifting shaft is driven by the primary motor through cog belt. Two slide gears whose tooth numbers are 29 (for low speed transmission) and 84 (for high speed) respectively are mounted on the shaft. Both gears and a bush (with inner spline) are fixed together as an integral, and can axially slide along the spline that is machined on the surface of the shaft. A hydraulic fork is used to perform gear selecting. The left end of the shifting shaft is supported by ball bearings with fixed outer ring, and the right end is supported by roller bearings with unfixed outer ring to consume heat expansion.

如图 4-14 所示，变速轴（轴Ⅰ）是花键轴，通过同步带由主电动机驱动。轴上花键部分安装有双联滑移齿轮，齿数分别为 29（低速用）和 84（高速用）。双联滑移齿轮为分体组合形式，由液压缸带动拨叉，使变速轴沿轴Ⅰ上的花键作轴向移动进行挡位选择。变速轴左端由球轴承支承，外圈固定，而右端由圆柱滚子轴承支承，外圈在箱体上不固定，以降低热变形的影响。

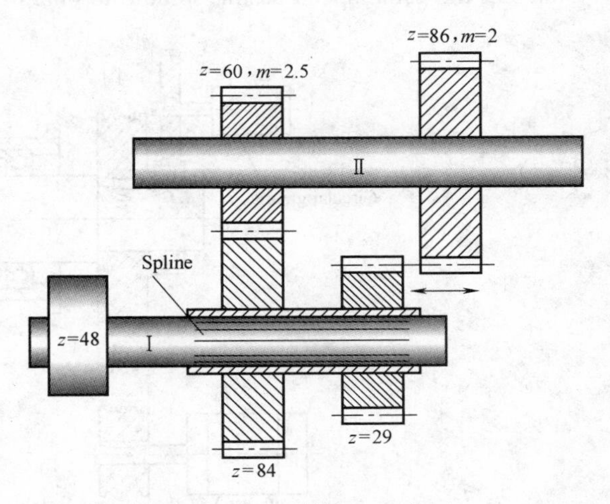

Figure 4-14 Gear shifting mechanism

(3) Spindle components (see Figure 4-15) The spindle is a hollow step shaft (usually made of 35CrMo). The front end of the spindle is a flange with conical boss for chuck positioning and mounting. The cylindrical hole that is axially through the spindle is used to accommodate those workpieces that which are large in length, and can also be used to mount hydraulic, pneumatic or electric clamping mechanisms.

The spindle is required for high speed and rigidity, therefore, both ends of the spindle are supported by angular contact bearings to withstand both axial and radial loads.

Adjusting nuts are used for bearing preloading set.

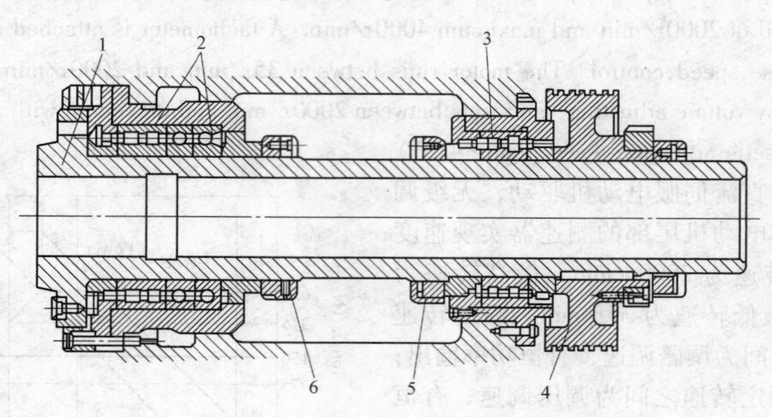

Figure 4-15 Spindle components of TND360

1—Spindle 2—Front bearing 3—Rear bearing 4—Pulley 5、6—Adjusting nuts

数控机床的主轴是一个空心的阶梯轴。主轴前端采用短圆锥法兰盘式结构，用于定位安装卡盘。主轴内孔用于通过长的棒料及卸下顶尖时穿过钢棒，也可用于通过气动、电动及液压夹紧装置的机构。

主轴安装在两个轴承上，这种主轴转速较高，要求的刚性也较高。所以前后轴承都用角接触球轴承（可以承受径向力和轴向力）。

(4) Circular grating The gear mounted on one end of the circular grating shaft has 60 teeth. It is driven by another 60-tooth gear that attached on the spindle. Therefore, the circular grating rotates at the same speed as the spindle to obtain spindle rotation speed.

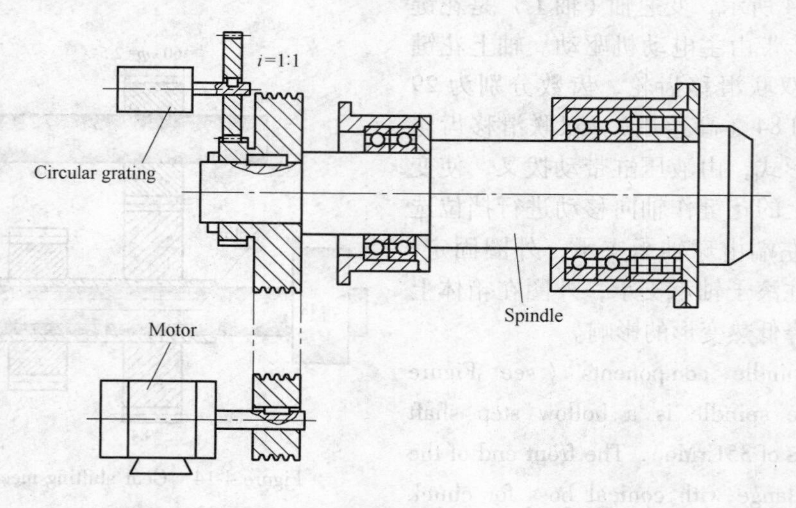

Figure 4-16 Spindle tachometer shaft of CNC lathe

(5) Spindle case The spindle case is made of cast iron, and is assembled by bolts or welding.

4.2.2 Feeding transmission

Feeding transmission of the TND360 CNC lathe consists of *Z*-direction and *X*-direction

transmissions, both of which are driven by DC servomotors.

Since the dimension and contour accuracy of workpieces depend on accuracy, sensitivity and stability of transmission system, servomotors for feeding drives have more stringent demands than that for primary drive.

To meet the requirements of machining accuracy, and to improve transmission accuracy and system rigidity, transmission system of the CNC lathe has minimized friction and inertia of moving parts. Clearance eliminating mechanisms are also necessary.

TND360数控车床进给传动系统由纵向进给传动装置（Z向）和横向进给传动装置（X向）组成，均由直流伺服电动机直接驱动。进给传动系统电动机要比主传动系统电动机的要求高，因为工件最后的尺寸精度和轮廓精度都直接受进给运动的传动精度、灵敏度和稳定性的影响。因此，对数控车床的进给传动系统，应特别注意减小摩擦力，提高传动精度和刚度，消除传动间隙以及减少运动件的惯量等。

1. Longitudinal feeding transmission

The longitudinal (*Z*-direction) feeding transmission consists of a carriage and a ball screw-nut system. The carriage is supported by bed ways, one of which has a triangular section and the other square. Both the ways are covered with low friction plastic layer, who has almost the same values of dynamic and static coefficients of friction.

The holders (压板) are mounted on the carriage to prevent the carriage from toppling under cutting force. Rubber shields are applied on both ends of the carriage to keep the contact surfaces clean for less abrasion.

纵向进给传动系统主要由纵向滑板和纵向滚珠丝杠副组成，它可以沿床身导轨作纵向移动。导轨的截面形状是三角形和平面的组合。在滑动导轨表面覆有一层摩擦系数极小的塑料涂层，动静摩擦系数接近。为了防止由于切削力作用而使滑板颠覆，纵向滑板的前后装有压板。滑板导轨部分的端面装有橡胶挡板，用来清除床身导轨表面上的切屑、灰尘等杂物，以减少导轨的磨损。

Figure 4-17 shows a typical longitudinal feeding transmission of a CNC lathe. The DC servomotor drives the ball screw-nut system through the cog belt to realize the *Z*-direction feeding motion of the carriage. Pulse equivalent of the transmission is 0.001mm.

The transmission system is under semi-closed loop control since the pulse encoder counts numbers of revolutions rather than detects the real carriage position and the motion speed.

The motor shaft and the cog pulley are connected by conical rings.

图4-17所示为MJ-50数控车床纵向进给传动系统。直流伺服电动机经同步带驱动滚珠丝杠螺母副，带动纵向滑板沿床身运动。纵向进给运动脉冲当量为0.001mm。纵向滑板的运动为半闭环控制，其位移量和移动速度并非直接测得，而是通过脉冲编码器将丝杠的旋转角度、转速信息反馈到数控系统进行转换，间接获得。

电动机轴与带轮通过无键锥环联轴器进行联接，以消除反向间隙。

Some CNC lathes apply safety couplings between the servo motor and the driven shaft to protect transmission components from excessive feeding force or carriage overloading.

When the torque is too large, the right-half disc of the coupling slips, and then the torque

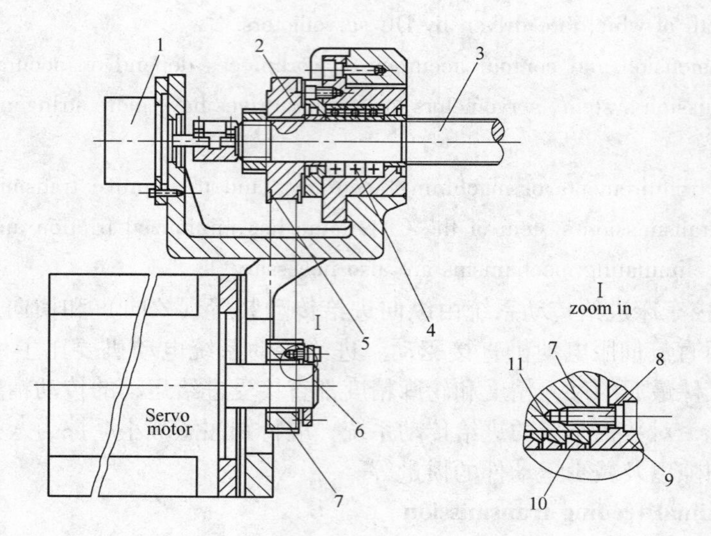

Figure 4-17　Longitudinal feeding transmission of CNC lathe

1、6—Pulse encoders　2、7—Cog pulleys　3—Ball screw　4—Bearings

5—Adjusting nut　8—Bolt　9—Flange　10、11—Conical rings

cannot be delivered. The slipping sends PLC an overloading signal to perform braking and alarming through CNC device (see Figure 4-18).

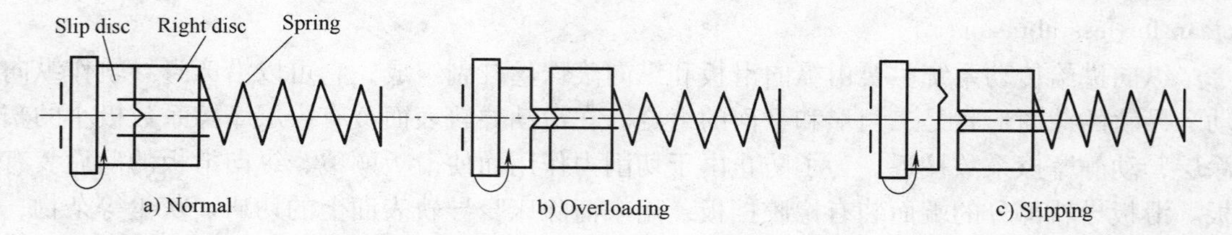

Figure 4-18　Work principle of safety coupling

Permissible transmission load depends on spring force. It can be conveniently set by an adjusting nut.

有些数控车床的进给传动系统中采用安全联轴器，其作用是当进给力过大或滑板过载时终止运动的传递，以避免整个传动机构的损坏。过载时，滚珠丝杠上的转矩增大，使联轴器的右半部分被推开，从而产生打滑现象，将传动链断开。同时，传感器产生过载报警信号，通过机床 PLC 使进给系统制动并报警，如图 4-18 所示。弹簧的预紧力可通过其上的调整螺母调整。

2. Traverse feeding transmission

The traverse carriage is mounted on the ways of the the longitudinal carriage to perform motion along x-axis. The work principle of traverse feeding transmission (see Figure 4-19) is similar to that of longitudinal feeding.

For slant longitudinal carriages, brake system is required.

横向滑板通过导轨安装在纵向滑板上，其传动原理与纵向滑板类似。对于倾斜床身，横向传动系统必须有制动装置，以防滑落。

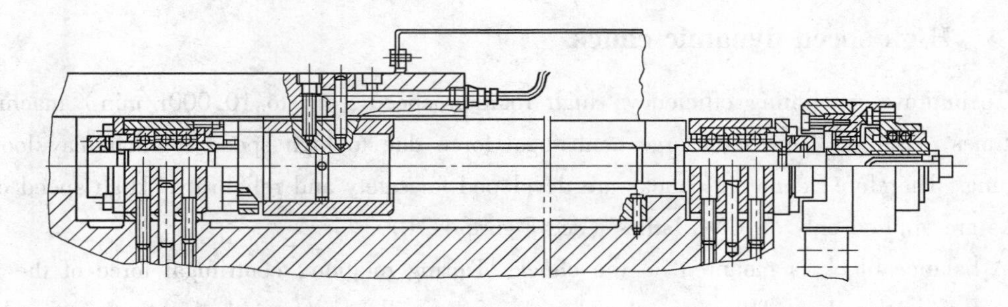

Figure 4-19 Traverse feeding transmission of TND360

4.2.3 Tailstock

The tailstock is mounted on the bed ways.

As Figure 4-20 shows, the conical hole of the sleeve (2) connects and drives the center (1). The other end of the sleeve connects the piston (4), which is actuated by electromagnetic valve that is under CNC device control. When hydraulic oil is injected into the left cavity of the cylinder, the sleeve extends with the center. (Notice that here is a static piston with a movable cylinder.) Once the limit switch (8) is pressed by the adjustable limiter (5), the center stops. The travel distance can be set by the position of the adjustable limiter. Then the hydraulic system injects oil into right cavity to withdraw the sleeve. The stroke limiter (6) touching the switch (7) means the sleeve is back to its origin, and then a signal will be sent to the CNC device to stop the motion.

数控车床的尾座安装在床身导轨上，其位置可根据工件的长短进行调整。

顶尖1与尾座套筒2用锥孔连接，尾座套筒可带动顶尖一起移动。在机床自动工作循环中，机床数控系统可根据加工程序控制尾座套筒的移动。当数控系统发出尾座套筒伸出的指令后，液压电磁阀动作，压力油通过活塞杆4的内孔进入尾座套筒2的左腔，推动尾座套筒伸出。当移动挡块5压下确认开关8，尾座套筒停止运动。尾座套筒移动的行程通过调节移动挡块5的位置来完成。当数控系统控制液压系统使压力油进入尾座套筒的右腔，尾座套筒退回。当固定挡块6压下确认开关7，数控系统即收到套筒退回到位的信号，套筒停止运动，如图4-20所示。

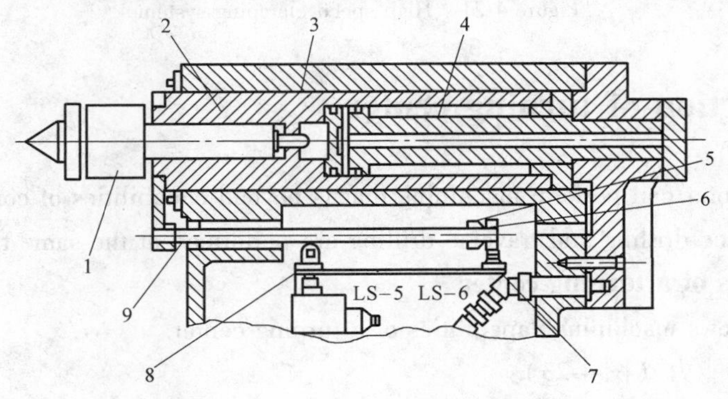

Figure 4-20 Construction of standard tailstock

1—Center 2—Sleeve 3—Body 4—Piston bar 5—Adjustable limiter

6—Fixed limiter 7、8—Limiter switch 9—Limiter bar

4.2.4 High speed dynamic chuck

To improve machining efficiency, high rotation speed (up to 10,000r/min) machining is sometimes required. However, large centrifugal force due to high speed rotation may loosen the clamping. Therefore, dynamic chucks are developed for safety and reliability. High speed dynamic chucks are applied only on CNC lathes (see Figure 4-21).

A balance block is mounted on the chuck. During rotation, centrifugal force of the balance weight presses the claw. Therefore, the clamping force will increase with the rotation speed rising. For the same rotation speed, increasing the mass of the balance can also produce larger clamping force.

在数控机床中，高速动力卡盘（图 4-21）一般只用于数控车床。为提高数控车床的生产效率，对主轴转速的要求越来越高，有的数控车床最高转速甚至达到10000r/min。对于这样高的转速，卡爪组件产生的离心力会使夹紧力明显减小，因此必须采用高速动力卡盘才能保证加工的安全性和可靠性。

高速回转时，卡爪上平衡块所产生巨大的离心力将变成压向卡爪座的夹紧力，转速越高，夹紧力越大。转速相同的情况下，增加平衡块的质量也可以相应增大夹紧力。

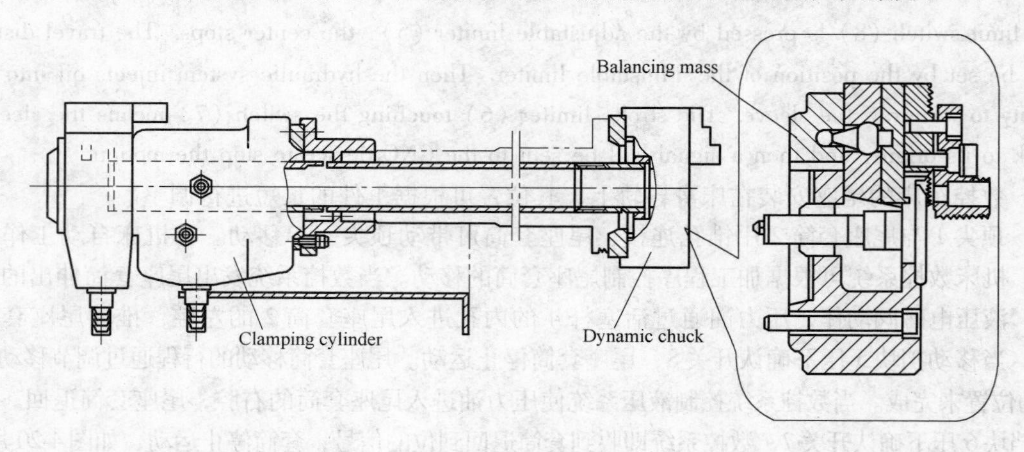

Figure 4-21 High speed clamping system

4.3 Introduction of turning center

The turning center can perform machining that is beyond capabilities of conventional lathes. In Figure 4-22, end face drilling and traverse drilling are performed at the same time.

1. Capabilities of a turning center

Figure 4-23 shows machining capabilities of a turning center.

车削中心工艺能力（图 4-23）：

① 铣端面槽，如图 4-23a 所示。

② 铣扁方，如图 4-23b 所示。

③ 端面钻孔、攻螺纹，如图 4-23c、d 所示。

Figure 4-22　Simultaneous holes making on the end and cylindrical surface of a shaft

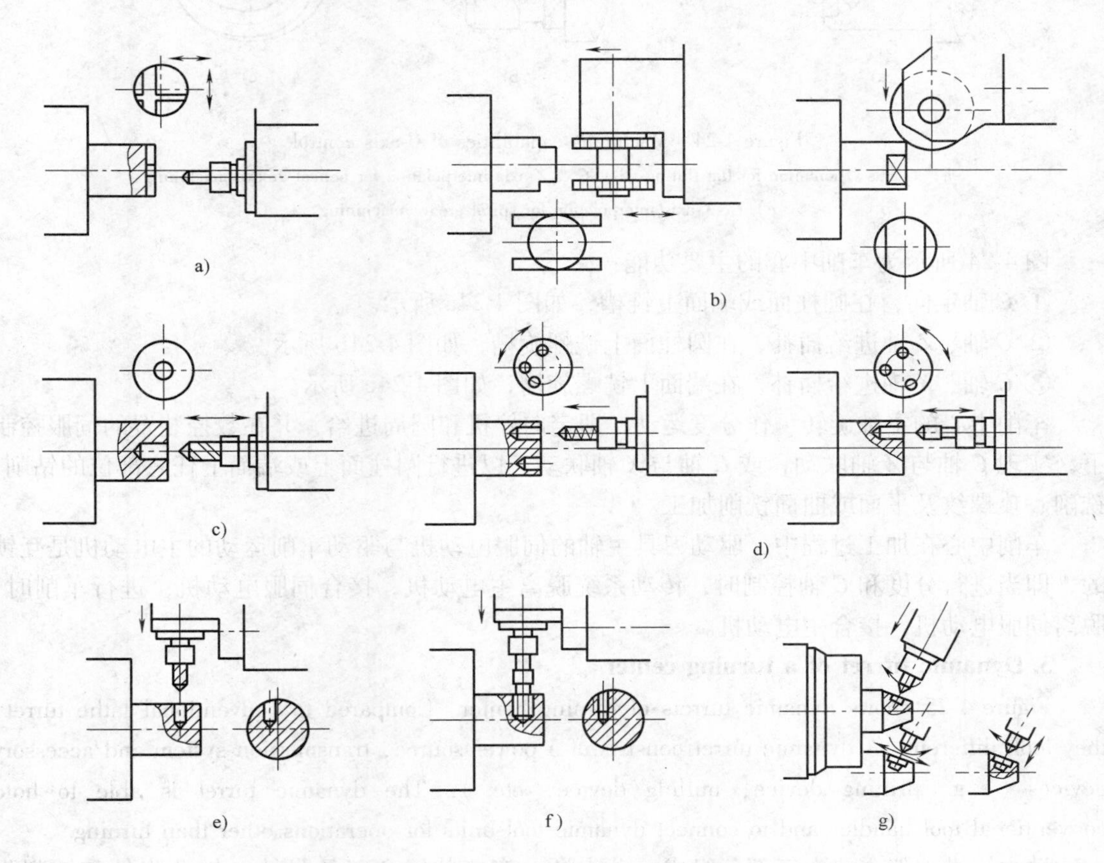

Figure 4-23　Machining capabilities of a turning center

a) Face fluting　b) Flat surface milling　c) Drilling on face

d) Indexing drilling & tapping　e) Transverse drilling　f) Transverse tapping　g) Conical surface drilling

④ 端面分度钻孔、攻螺纹。每加工完一孔由主轴带动工件分度。

⑤ 横向钻孔、横向攻螺纹和斜面钻孔、攻螺纹，如图 4-23e、f、g 所示。

2. *C*-axis of a turning center

Besides primary motion of turning, the spindle can perform indexing motion, i. e. orientation and circular feeding.

Figure 4-24 shows machining capabilities of *C*-axis control. Interpolation of *C* & *Z*-axis, or *C* & *X*-axis can machine helical or spiral grooves. The spindle motor drives the primary motion and the servo motor for *C*-axis feeding. The spindle motor is locked during indexing and *C*-axis control, and the servo motor is locked during turning operation. The motors are not allowed to run simultaneously.

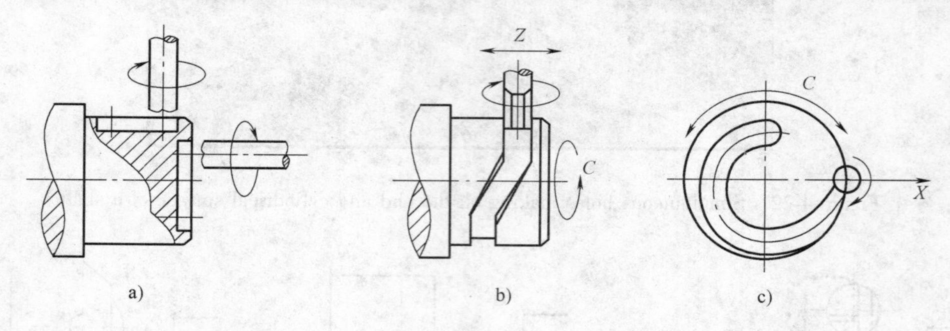

Figure 4-24　Machining capabilities of *C*-axis control

a) *C*-axis Orientation for flat fluting　b) *C* & *Z*-axis interpolation for helical groove machining

c) *C*&*X*-axis interpolation for spiral groove machining

图 4-24 所示为车削中心的主要功能：

① *C* 轴定向，在圆柱面或端面上铣槽，如图 4-24a 所示。

② *C* 轴、*Z* 轴进给插补，在圆柱面上铣螺旋槽，如图 4-24b 所示。

③ *C* 轴、*X* 轴进给插补，在端面上铣螺旋槽，如图 4-24c 所示。

车削中心的主轴旋转可作分度运动，即定向停机和圆周进给，并在数控装置的伺服控制下，实现 *C* 轴与 *Z* 轴联动，或 *C* 轴与 *X* 轴联动，以进行圆柱面上或端面上任意部位的钻削、铣削、攻螺纹及平面或曲面铣削加工。

车削中心在加工过程中，驱动刀具主轴的伺服电动机与驱动车削运动的主电动机是互锁的，即当进行分度和 *C* 轴控制时，传动系统脱离主电动机，接合伺服电动机；进行车削时，脱离伺服电动机，接合主电动机。

3. Dynamic turret of a turning center

Figure 4-25 shows dynamic turrets of turning center. Compared to conventional lathe turret, they look different. A dynamic turret consists of a power source, transmission system and accessory devices (e. g. drilling device, milling device, etc.). The dynamic turret is able to hold conventional tool handles and to connect dynamic tool units for operations other than turning.

动力转塔刀架主要由三部分组成：动力源、变速装置和刀具附件（钻孔附件和铣削附件等）。刀架上既可以安装各种普通刀具进行车削加工，还可安装动力刀具进行铣、钻、镗等加工，从而实现加工的自动化和高效化。

Figure 4-26 shows the construction of a high speed drilling unit. When in use, the sleeve (4)

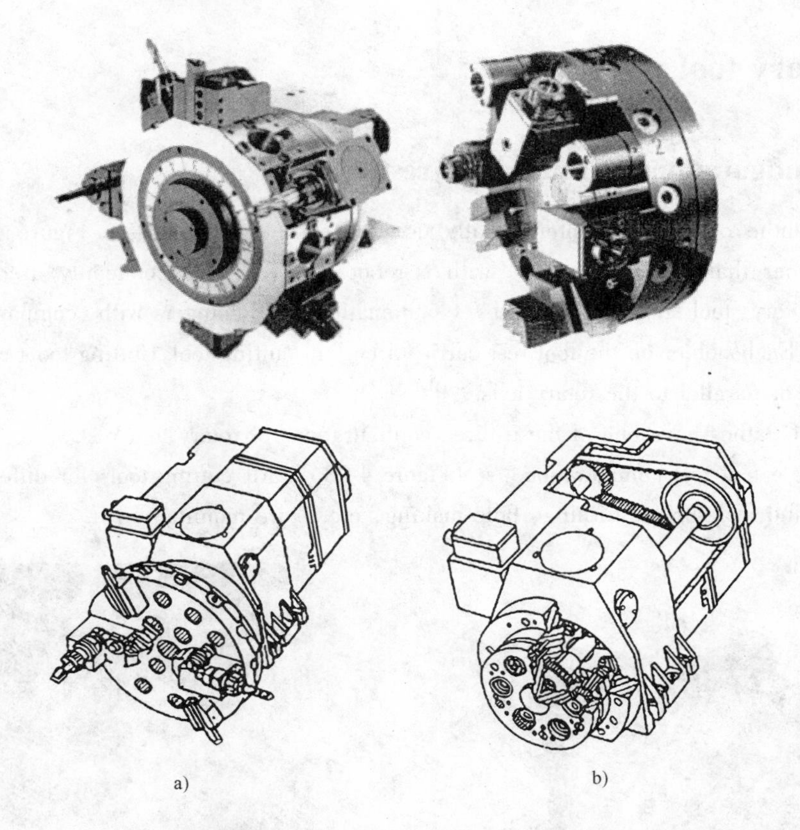

Figure 4-25 Dynamic turret of CNC lathe

a) Outsidew view b) Construction

is connected to the tool holder of dynamic turret. The bevel gear (1) on the right end of the main shaft (3) engages to the central driving bevel gear in turret. The drill bit is clamped by the spring connector locating on the left end of the main shaft.

图 4-26 所示为高速钻孔附件，轴套 4 的 *A* 部装入转塔刀架的刀具孔中。主轴 3 的右端装有锥齿轮 1，与动力转塔刀架的中央驱动锥齿轮啮合。钻头通过主轴头部的弹簧夹头 6 与主轴连接。

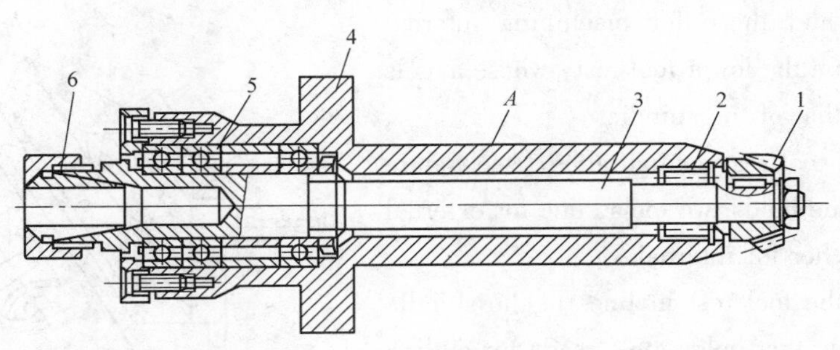

Figure 4-26 Construction of high speed drilling unit

1—Bevel gear 2—Needle bearing 3—Main shaft

4—Sleeve 5—Angular contact bearings 6—Spring connector

4.4 Rotary tool rest

4.4.1 Configurations of rotary tool rest

A CNC lathe or turning center usually uses a rotary tool rest (see Figure 4-27) for tool mounting, rather than a tool magazine with a robot arm which is commonly used on a milling center. The rotary tool rest is a kind of automatic tool changer with comparatively simple construction. Each holder on the tool rest carries a certain cutting tool. Cutting tools can be mounted perpendicular or parallel to the rotary axis.

Most CNC lathes have one of these three configurations of rotary tool rest.

(1) Single tool rest configuration (see Figure 4-28) All cutting tools for different machining purposes (cylindrical surface turning, hole making, etc.) are mounted on it.

Figure 4-27 A rotary tool rest for CNC lathe

Figure 4-28 Single tool rest configuration

(2) Double tool rest configuration Cutting tools for machining external surface are mounted on the upper tool rest, whose axis is parallel to that of the spindle; and those for machining internal surface are fixed on the lower tool rest, whose axis is perpendicular to that of the spindle.

(3) Conical tool rest configuration (see Figure 4-29) Each plane holds two tools, one for external surface and the other for internal.

The axis of the tool rest inclines to the spindle axis, but the plane that holds tools ready for cutting must be vertical to the spindle axis.

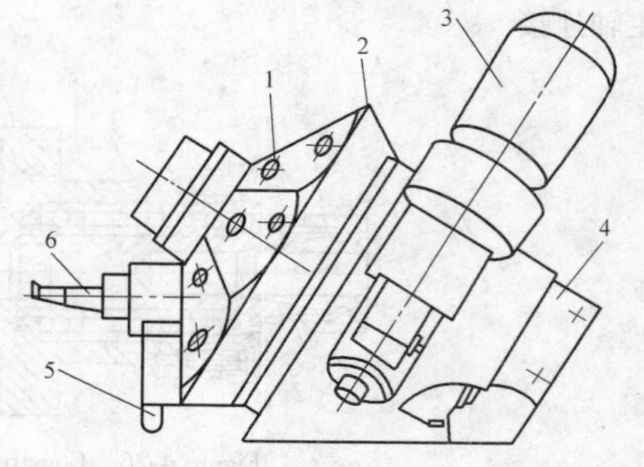

Figure 4-29 Conical tool rest configuration

1—Tool holders 2—Turret head 3—Motor 4—Base
5—Tool for external surface 6—Tool for internal surface

The more holders does a tool rest have, the narrower angle between adjacent cutting tools, which

may result in interference between the tools that are not in use and the workpiece. Therefore, a rotary tool rest may have up to 20 holders, but 8-, 10-, 12-, and 16-holder are popular.

For accuracy, a rotary tool rest must have sufficient strength and rigidity to withstand cutting force during rough machining, and minimize tool rest distortion for accuracy. Besides, reliable positioning mechanism is required to ensure repositioning accuracy after tool changing.

在数控车床上，回转刀架和其上的刀具布置大致有以下几种类型：

（1）一个回转刀架　外圆类、内孔类刀具混合放置（图4-28）。

（2）两个回转刀架　上刀架的回转轴与主轴平行，用于装外圆类刀具；下刀架的回转轴与主轴垂直，用于装内孔类刀具。

（3）双排回转刀架　外圆类、内孔类刀具分别布置在刀架的一侧面（图4-29）。

回转刀架的工位数最多可达20多个，但最常用的是8、10、12和16工位4种。工位数越多，刀间夹角越小，非加工位置刀具越容易与工件产生干涉。

回转刀架在结构上必须具有良好的强度和刚度，以承受粗加工时的切削力，应减小刀架在切削力作用下的位移变形，提高加工精度。回转刀架还应定位可靠，以保证每次换刀之后的重复定位精度。

4.4.2 Tool changing procedure of rotary tool rest

Figure 4-30 shows the construction of the rotary tool rest of the CK3263.

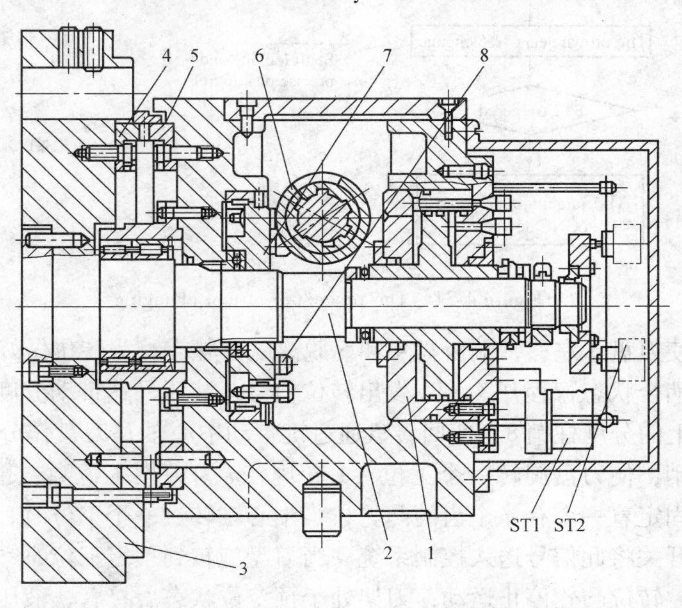

Figure 4-30 Construction of rotary tool rest of CK3263

1—Cylinder 2—Shaft 3—Tool rest 4、5—Crown gear

6—Indexing cam 7—Wheel 8—Indexing pins ST1—Counter ST2—Switch

The procedure of tool changing can be illustrated by Figure 4-31.

CK3263系列数控车床回转刀架结构如图4-30所示，回转刀架的升起、转位、夹紧等动作均由液压驱动。当数控装置发出换刀指令以后，液压油进入液压缸1的右腔，通过活塞推

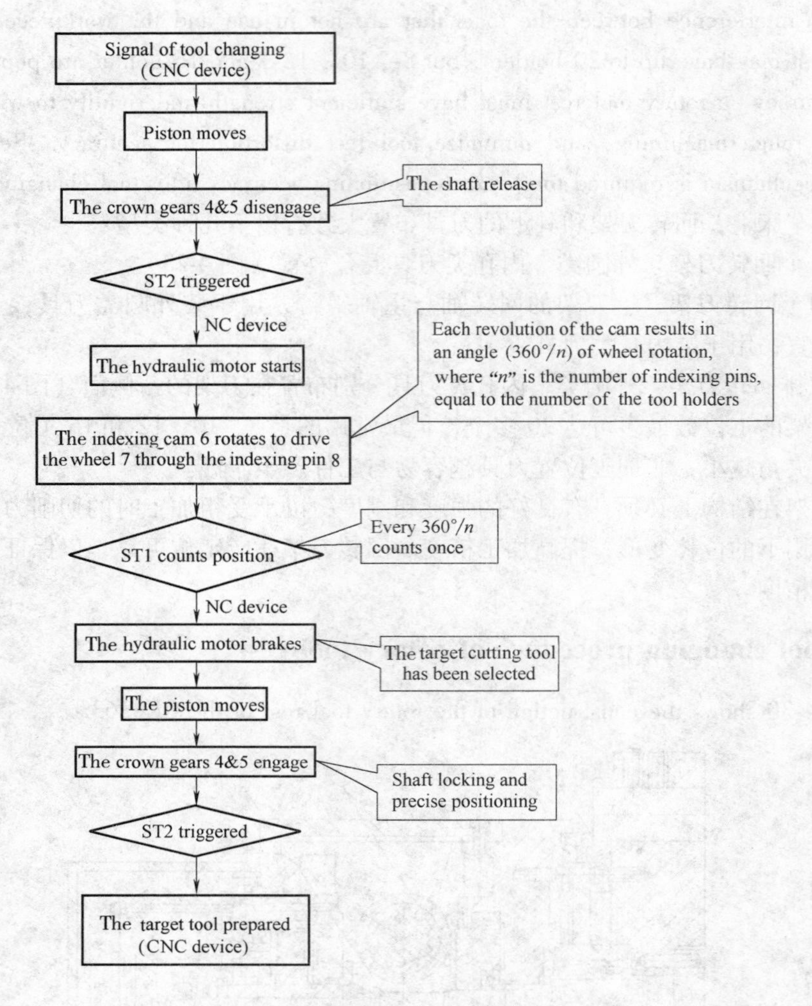

Figure 4-31　The procedure of tool changing

动刀架中心轴2使刀盘3左移，端齿盘4和5脱离啮合状态，为转位做好准备。齿盘处于完全脱开位置时，啮合状态行程开关ST2发出转位信号，液压马达带动分度凸轮6旋转，凸轮依次推动转盘7上的分度柱销8，使回转盘通过键带动中心轴及刀盘作分度转动。凸轮每转一周拨过一个柱销，使刀盘旋转一个工位（$1/n$ 周，n 为刀架工位数，也等于柱销数）。刀架中心轴的尾端固定着一个有 n 个齿的凸轮，当中心轴转过一个工位时，凸轮压合计数行程开关ST1一次，开关将此信号送入控制系统。当刀盘旋转到预定工位时，控制系统发出信号使液压马达制动，转位凸轮停止运动，刀架处于预定位状态。接着，液压缸1左腔进油，通过活塞将刀架中心轴和刀盘拉回，使端齿盘啮合，刀盘完成精定位和夹紧。此时刀架中心轴尾部将ST2压下，发出转位结束信号。图4-31所示为回转刀架换刀步骤。

4.4.3　Principle of tool rest indexing

In Figure 4-32, during the period that the cylindrical cam turns 180° from the present position, the pin B engages with the groove on the cam and moves leftwards (thus turning the wheel 3), and

finally reaches to the position where the pin *A* was. At the same time, the pin *C* moves to the position where the pin *B* was, and the pin *A* disengages with the cam. Cam reversing can also drive the wheel, because CNC device will decide the rotational direction for the shortest time consumption of tool changing.

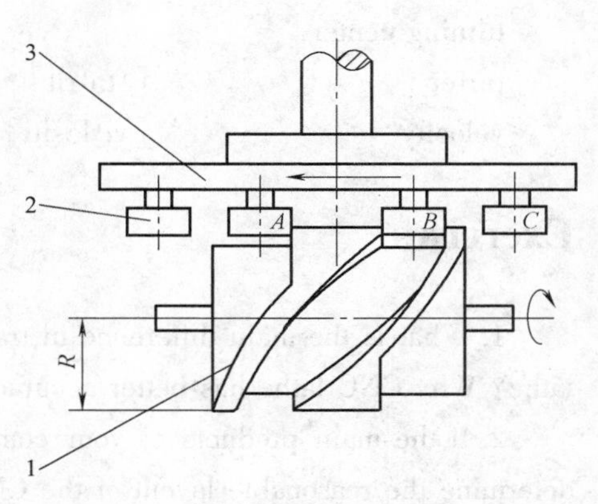

Figure 4-32 Cylindrical cam indexing

1—Indexing cam 2—Indexing pin 3—Wheel

刀盘转位驱动采用圆柱凸轮步进传动机构，其工作原理如图 4-32 所示。从动回转盘端面有多个柱销，数量与工位数相等。当凸轮按图中所示方向旋转时，*B* 销先进入凸轮轮廓的曲线段，这时凸轮开始驱动回转盘转位，与此同时，*A* 销与凸轮轮廓脱开。当凸轮转过 180°时，*B* 销接触的凸轮轮廓由曲线段过渡到直线段，同时与 *B* 销相邻的 *C* 销开始与凸轮的直线轮廓的另一侧面接触，此时凸轮继续转动，但回转盘不动，刀架处于预定位状态。凸轮正反转时均可带动回转盘作正反方向的旋转。

Glossary

carriage	[ˈkæridʒ]	（机床的）滑板
chuck	[tʃʌk]	卡盘
classification	[ˌklæsifiˈkeiʃən]	分类；分级
cutting fluid		切削液
cylindrical cam		圆柱凸轮
disengage	[ˌdisinˈgeidʒ]	分开；使脱离
fluting	[ˈfluːtiŋ]	开槽
headstock	[ˈhedstɔk]	（车床等的）主轴箱
highlight	[ˈhailait]	使显著；强调
layout	[ˈleiaut]	布局
reaming	[riːmiŋ]	铰孔加工
respectively	[riˈspektivli]	分别地
revolutionary	[ˌrerəˈluːʃənəri]	回转的；旋转的
shield	[ˈʃiːld]	防盾；防护罩
simultaneous	[ˌsaiməlˈteinjəs]	同时发生的
slant	[slɑːnt]	倾斜的
slide	[slaid]	（机床）滑轨；滑座
sufficient	[səˈfiʃənt]	足够的；充分的
tailstock	[ˈteilstɔk]	（车床的）尾座
tapping	[tæpiŋ]	攻螺纹
tool post		刀架

turning center		车削中心
turret	[ˈtəːrit]	转台，转塔刀架
velocity	[viˈlɔsiti]	速度

Exercises

1. What is the main difference in transmission system between a conventional lathe and a CNC lathe? Why CNC lathe has better accuracy in terms of transmission?

2. If the main products of your company are large diameter iron housings ($d > 6,000$mm), determine the reasonable layout of the CNC lathe.

3. Give five examples of products that can be machined by a turning center.

4. When a turning center is drilling holes on cylindrical surface, which component is performing primary motion?

Chapter 5　CNC Milling Machine

5.1　Introduction of CNC milling machines

5.1.1　Machining capabilities of CNC milling machines

Milling is one of the most commonly used machining processes, including flat surface milling and contour milling. CNC milling machines are especially suitable for machining these surfaces:

(1) Flat surfaces (see Figure 5-1)　The angle between machining surface and horizontal is constant. It includes flat surfaces on a workpiece, or the surfaces that can be unfolded into flat surfaces (e. g. curved surfaces, inclined surfaces, conical surfaces, etc).

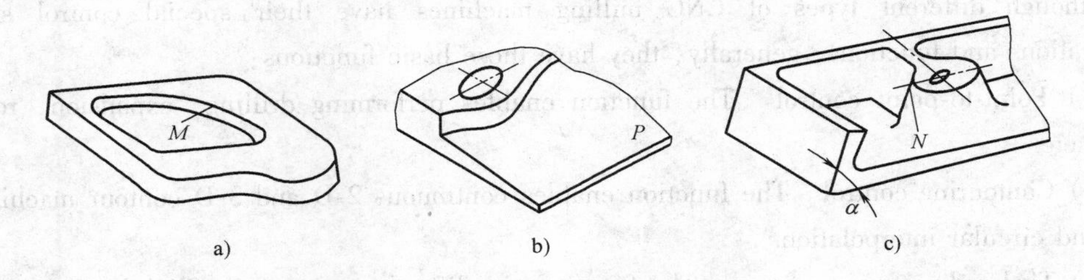

Figure 5-1　Examples of flat surfaces

a) Cylindrical surface　b) lnclined surface　c) Conicaal surface

平面类零件是指加工平面与水平面的夹角为一固定角的零件，如图5-1所示。这类零件的各加工表面为平面或可以展开为平面的曲面，如曲线轮廓面、圆锥面等。

(2) Torsional surfaces (see Figure 5-2)　The angle between the machining surface and the horizontal is continuous varying. A torsional surface cannot be unfolded into a flat surface, but the contact between the rotary surface of the milling tool and the workpiece is always a straight line at any moment.

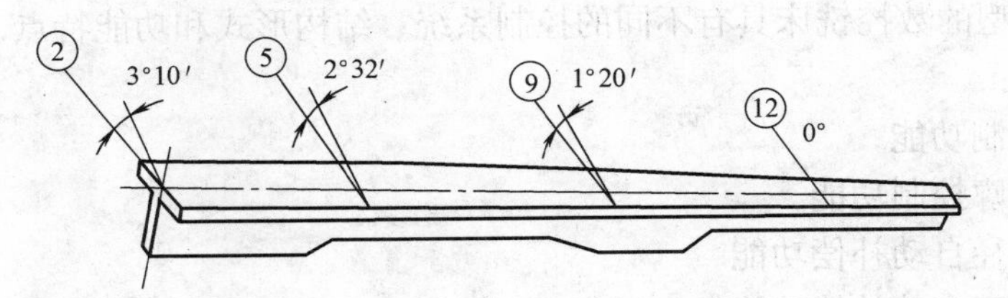

Figure 5-2　An example of torsional surface

变斜角类零件是指加工面与水平面的夹角为连续变化的零件，如图5-2所示。这类零件的加工面不能展开为平面，但在加工中，铣刀圆周与加工面的接触为一直线。

(3) Curved surface (see Figure 5-3) (3-D contour) A 3-D contour cannot be unfolded into a flat surface, and the contact between the milling tool and the workpiece is always a point.

空间曲面不能展开为平面，且加工面与铣刀始终为点接触，如图 5-3 所示。

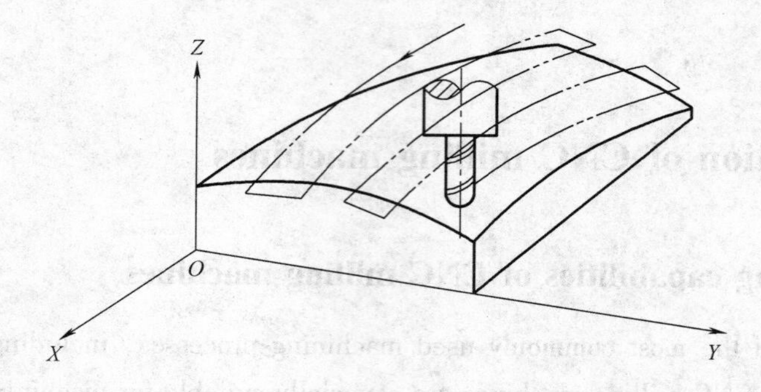

Figure 5-3 An example of 3-D contour

5.1.2 Main functions of CNC milling machines

Although different types of CNC milling machines have their special control systems, configurations and functions, generally, they have these basic functions:

(1) Point-to-point control The function enables performing drilling, expansion, reaming, boring, etc.

(2) Contouring control The function enables continuous 2-D and 3-D contour machining by linear and circular interpolation.

(3) Tool radius compensation The function simplifies programming. Slightly changing offset value can conveniently machining fitting pairs (molds) with the same program.

(4) Tool length compensation With this function, surface cutting depth can be easily adjusted thus less requirement of tool length accuracy.

(5) Mirror machining For the symmetrical workpiece, machining can be performed by programming for only a half.

(6) Fixed cycle It simplifies programming for those repeatedly basic motions. The principle of the function is a built-in subprogram.

(7) Other special functions e. g. profile modeling.

虽然不同类型的数控铣床具有不同的控制系统、结构形式和功能特点，但它们具有一些共同的基本功能：

(1) 点位控制功能

(2) 连续轮廓控制功能

(3) 刀具半径自动补偿功能

(4) 刀具长度自动补偿功能

(5) 镜像加工功能

(6) 固定循环功能

(7) 特殊功能，如数控仿形加工等

5.2 Layouts and types

5.2.1 Layouts determined by weight and dimension of workpieces

Forming motion is the relative motion between cutting tool and workpiece, therefore, the motion distribution of CNC milling machines depends on the weight and dimension of the workpiece that can be machined. Figure 5-4 shows three layouts of CNC milling machines.

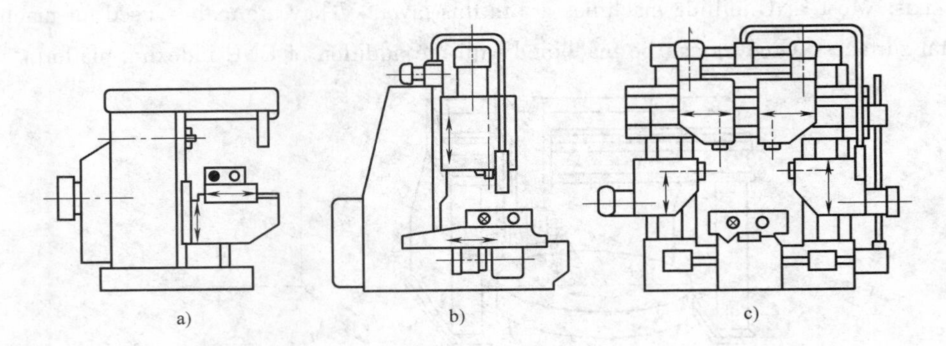

Figure 5-4 Layouts of CNC milling machines

a) Worktable feeding (Horizontal configuration) b) Milling head feeding in Z-axis (Vertical configuration)

c) Worktable feeding in Y-axis (Gantry configuration)

成形加工工件的运动实际上是刀具与工件的相对运动，因此，数控铣床的运动分配可以有多种方案。同是用于铣削加工的铣床，根据工件的重量和尺寸的不同，可以有不同的布局方案。

图 5-4 所示为数控铣床三种不同的总体布局示意图：

1）工作台作进给运动的卧式升降台铣床，如图 5-4a 所示。

2）铣头沿 Z 轴作垂直进给运动的立式升降台铣床，如图 5-4b 所示。

3）工作台沿 Y 轴作进给运动的龙门式数控铣床，如图 5-4c 所示。

5.2.2 Motion distribution and components layout

The number of feeding motions determines machining functions of a CNC milling machine. Therefore, motion distribution and components layout are important considerations during design stage.

For CNC boring-milling machine, four feeding components are required. The layout depends on the primary machining position.

If the top surface is the primary machining object, a vertical spindle is preferred to machine the surface, and also holes, bosses, threads, or slots on the surface by clamping once. Besides linear coordinates, a rotary/indexing platform can be added to expand machining capabilities.

If side surfaces are primary, a horizontal spindle is preferred. With a CNC rotary platform, many machining processes (e. g. milling, drilling, reaming, boring, tapping) on all sides can be performed by clamping once.

运动的分配与部件的布局是铣床总布局的中心问题。例如，如果需要对工件的顶面进行加工，则铣床主轴应布局成立式的。如果需要对工件的多个侧面进行加工，则主轴应布局成卧式的，并在三个直线进给坐标之外再加一个数控转台，以便一次装夹集中完成多面的铣、镗、钻、铰、攻螺纹等多工序加工。

5.2.3 Types of CNC milling machines

1. Classified by spindle positions and general layouts

(1) Vertical CNC milling machines (see Figure 5-5) The name comes from the vertical spindle axis. Most CNC milling machines are in this layout. They are mainly used for machining on horizontal surfaces. Grooves can be machined with the addition of CNC indexing platform.

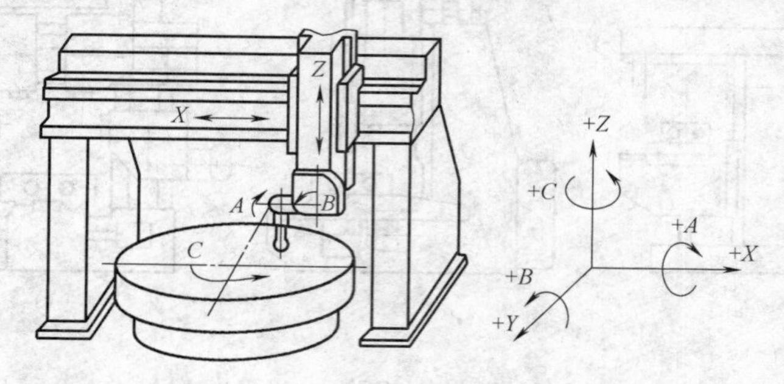

Figure 5-5 Vertical CNC milling machines

(2) Horizontal CNC milling machine (see Figure 5-6) With a horizontal spindle, a horizontal CNC milling machine is mainly used for machining on vertical surfaces. Equipped with a CNC rotary platform, the horizontal CNC milling machine can perform continuous rotary contouring or all sides machining by clamping once.

(3) Dual-position CNC milling machine (see Figure 5-7) The spindle axis can shift between horizontal and vertical positions. These milling machines can perform machining that can be done by either vertical or horizontal CNC milling machines.

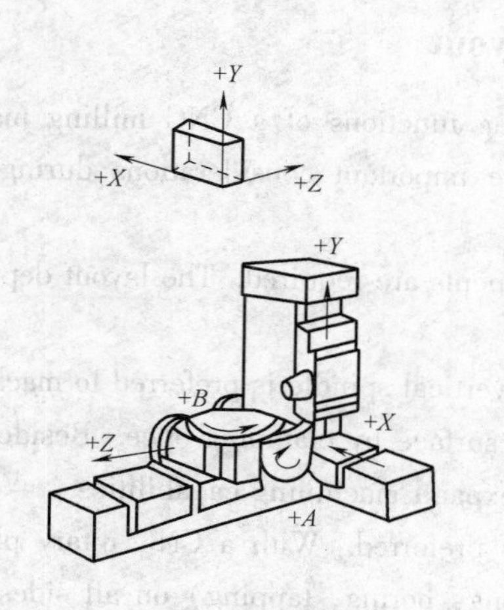

Figure 5-6 Horizontal CNC milling machine

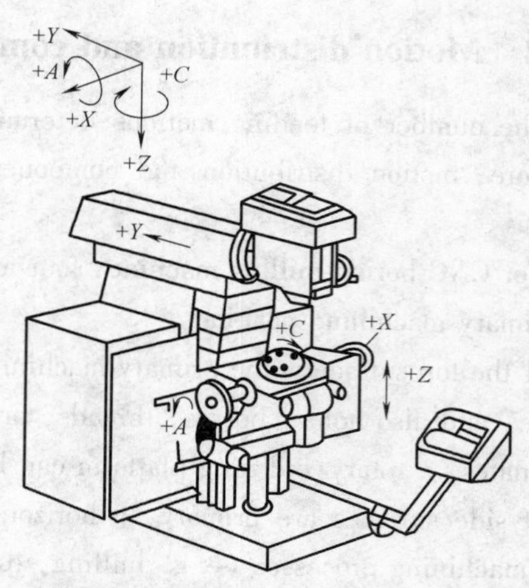

Figure 5-7 Dual-position CNC milling machine

With the addition of a universal spindle (the angle between the spindle axis and the horizontal can be any value), surfaces at any angle can easily be machined.

按机床主轴的布置形式及机床的布局特点，数控铣床通常可分为立式、卧式和立卧两用式三种。

(1) 立式数控铣床（图 5-5） 主轴轴线垂直于水平面，是数控铣床中最常见的一种布局形式，主要用于水平面内的型面加工，增加数控分度头后，可在圆柱表面上加工曲线沟槽。

(2) 卧式数控铣床（图 5-6） 主轴轴线平行于水平面，主要用于垂直平面内的各种型面加工，配置万能数控转盘后，还可以对工件侧面上的连续回转轮廓进行加工，并能在一次安装后加工箱体零件的各个侧面。

(3) 立卧两用式数控铣床（图 5-7） 主轴轴线方向可以变换，既可以进行立式加工，又可以进行卧式加工，使用范围更大，功能更强。若采用数控万能主轴（主轴头可以任意转换方向），就可以加工出与水平面成各种角度的工件表面。

2. Classified by functions of CNC systems

(1) Economy CNC milling machines (see Figure 5-8) They are built on a conventional milling machine with transmission modifications and simple CNC system. Economy CNC milling machines have low cost but less automation, less function and less accuracy. They are usually two-axis controlled (*X-Y*) for flat curve machining.

简易型数控铣床是在普通铣床的基础上，对机床的机械传动结构进行简单的改造，并增加简易数控系统后形成的。这种数控铣床成本较低，自动化程度和功能都较差，一般只有 *X*、*Y* 两坐标联动功能，加工精度也不高，可以加工平面曲线类和平面型腔类零件。

(2) Standard CNC milling machines (see Figure 5-9) They are professionally designed. With three-axis control, complex curved surfaces and shells (e. g. molds, cams, connecting rods, etc) can be machined.

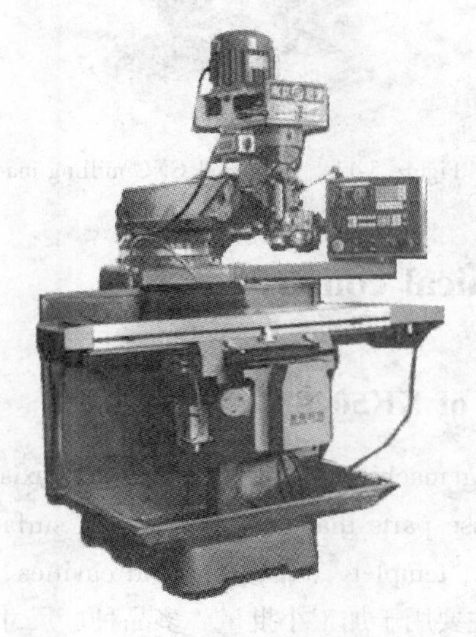

Figure 5-8 Economy CNC milling machine

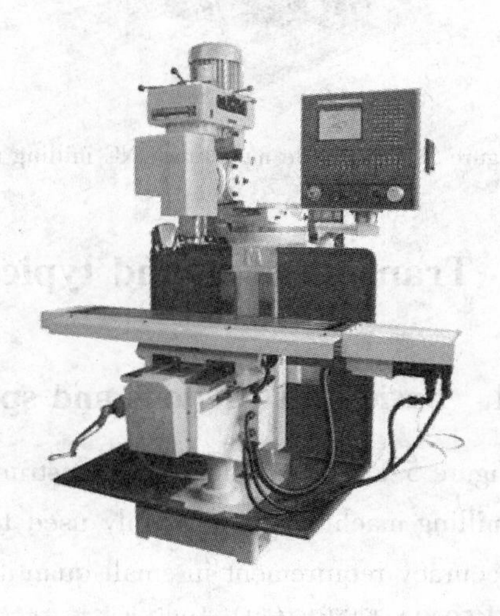

Figure 5-9 Standard CNC milling machine

普通数控铣床可以三坐标联动，用于各类复杂的平面、曲面和壳体类零件的加工，如各种模具、样板、凸轮和连杆等。

（3）Profile modeling CNC milling machines（see Figure 5-10） They are specially designed for complex mold cavities machining, especially for those irregular curved surfaces.

数控仿形铣床主要用于各种复杂型腔模具或工件的铣削加工，对由不规则的三维曲面和复杂边界构成的工件进行加工时更显示出其优越性。

（4）CNC universal milling machines（see Figure 5-11） Universal milling machines have the most freedom to machine parts with complex contours, e. g. cutting tools, fixtures. They are usually used to machine parts for repair rather than workpieces.

Equipped with CNC system and with modified transmissions, CNC universal milling machines have enhanced functions.

数控万能工具铣床在普通万能工具铣床的基础上，对机床的机械传动系统进行了改造，并增加了数控系统，从而使万能工具铣床的功能大大增强。这种铣床适用于各种工装、刀具对各类复杂的平面、曲面零件的加工。

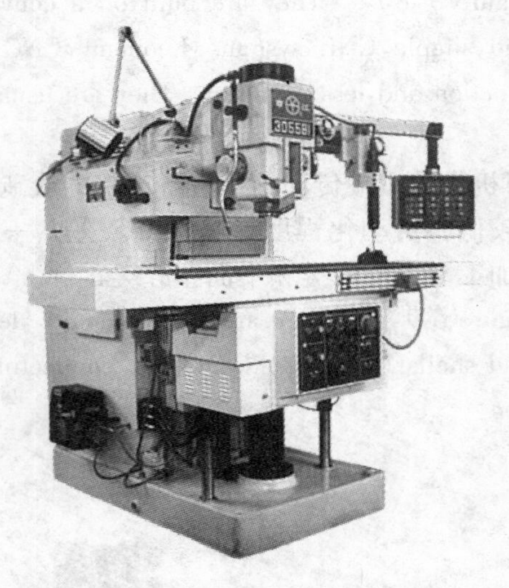

Figure 5-10 Profile modeling CNC milling machine

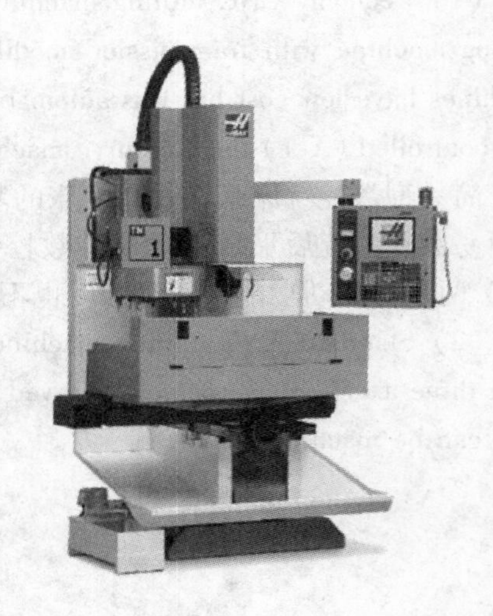

Figure 5-11 Universal CNC milling machine

5.3 Transmissions and typical mechanical constructions

5.3.1 Basic constructions and specifications of XK5040A

Figure 5-12 shows the basic construction of milling machine. XK5040A is a 2. 5-axis-control CNC milling machine. It is mainly used to machine those parts that have flat curved surfaces and high accuracy requirement in small quantity, e. g. cams, templets（靠模）, mold cavities, etc.

XK5040A 型数控铣床为两轴半控制的数控铣床，主要用于加工小批量、多品种、尺寸形状复杂、精度要求较高的零件，如凸轮、样板、靠模、模具弧形模等平面曲线，如图 5-12 所示。

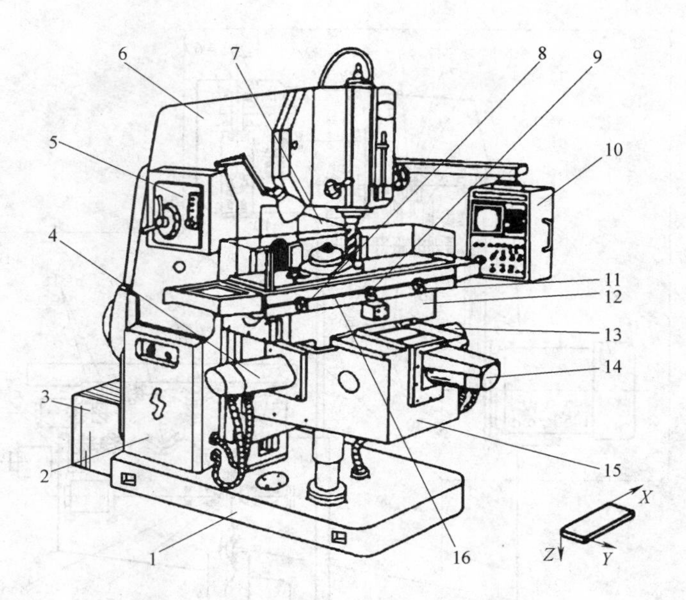

Figure 5-12 Basic construction of XK5040A

1— Base 2— Electrical cabinet 3—Transformer 4—Vertical feeding servo motor
5—Spindle speed shifting lever 6—Bed 7—CNC cabinet 8, 11— Safety switch
9— Stroke limiter 10—Control panel 12— Traverse carriage
13— Longitudinal feeding servo motor 14—Traverse feeding servo motor 15-Elevator 16—Worktable

5.3.2 Transmission of CNC milling machines

1. Primary transmission (see Figure 5-13, part A)

The primary motion (spindle rotation) of XK5040A is driven by an AC motor of 7.5kW through V-belt transmission (diameter of pulleysϕ140/ϕ285mm). By different combinations, three sliding gear pairs between shaft Ⅰ and shaft Ⅳ can provide eighteen reduction ratios. It makes the spindle shaft speed range between 30-1500r/min.

The bevel gears between shafts Ⅳ-Ⅴ and gears between shafts Ⅴ-Ⅵ are used to regulate the direction and positions of transmission between the shafts.

Spindle speed ranges 30-1500r/min. Figure 5-13 shows the transmission diagram of XK5040A.

XK5040A 型数控铣床的主运动是主轴的旋转运动，由 7.5kW 的主电动机驱动，经 ϕ140mm/ϕ285mmV 带传动，再经Ⅰ/Ⅱ轴、Ⅱ/Ⅲ轴、Ⅲ、Ⅳ轴间的滑移齿轮组变速，最后通过Ⅳ/Ⅴ轴间的锥齿轮副及Ⅴ/Ⅵ轴间的齿轮副传动，使主轴获得 18 级转速，转速范围为 30～1500r/min。图 5-13 为 XK5040A 型数控铣床的传动图。

2. Feeding transmissions (see Figure 5-13, part B)

Feeding transmission of XK5040A includes longitudinal, traverse and vertical directions.

(1) Longitudinal feeding (*X*-axis) It is driven by an FB-15 DC servo motor through clearance elimination gears ($z=18$) and ball screw-nut mechanisms.

(2) Traverse feeding (*Y*-axis) It has the same principle as that of the longitudinal feeding transmission.

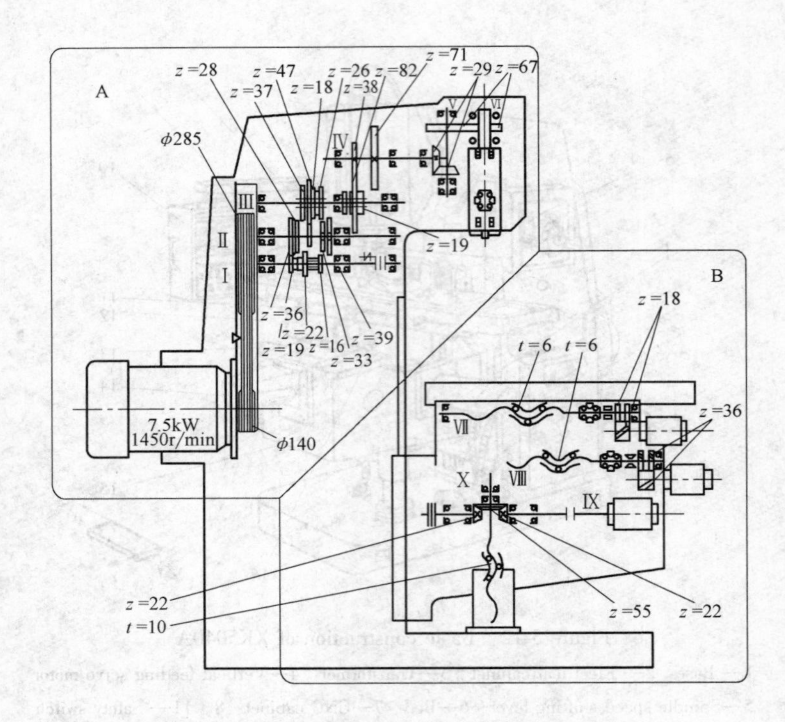

Figure 5-13　Transmission diagram of XK5040A

（3）Vertical feeding（*Z*-axis）　Another FB-15 DC servo motor drives the worktable through the bevel gears（$z=22$，$z=55$）and the ball screw-nut mechanism. Brake is equipped on the motor to avoid the worktable sliding down due to gravity when power off.

进给运动包括工作台的纵向、横向和垂直三个方向的进给运动。纵向进给由一台直流伺服电动机驱动，经斜齿轮副传动，带动滚珠丝杠转动，通过丝杠螺母机构实现。横向进给运动的传动原理与纵向进给传动相同。垂直进给运动也由一台直流伺服电动机驱动，经锥齿轮副传动，带动滚珠丝杠转动。垂直进给的伺服电动机带有制动器，以防止断电时升降台因自重而下滑。

5.3.3　Main mechanical components of CNC milling machines

1. Spindle

It is one of the most important components of a CNC milling machine. General requirements of the spindle include good accuracy, rigidity, vibration-resistance, thermal stability and wear resistance as the requirements of conventional milling machines, but the criteria is more stringent since no manual adjusting is expected during machining process.

Although a DC motor provides a wide spindle rotation speed range, a two-position gear shifting mechanism（see Figure 5-14）is often used to keep the motor operating at its most efficient rotation speed range.

主轴是数控铣床的重要组成部分，除了与普通铣床一样要求其具有良好的旋转精度、静刚度、抗振性、热稳定性及耐磨性外，由于数控铣床在加工过程中不进行人工调整，因此对

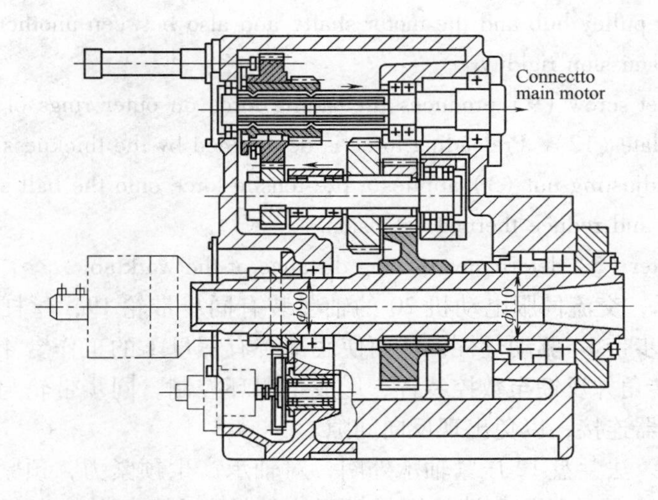

Figure 5-14　Two-position gear shifting mechanism

数控铣床主轴的要求更为严格。

图 5-14 所示为数控铣床典型的二级齿轮变速主轴结构。其主轴采用两支承结构，主电动机的运动经双联齿轮带动中间传动轴，再经一对圆柱齿轮带动主轴旋转。

2. Longitudinal transmission mechanism of the worktable

Figure 5-15 illustrates the longitudinal（*X*-axis）transmission components of the worktable. The servo motor（20）drives the worktable（4）through the cog belt transmission（11, 14, 19）and the ball screw-nut（1, 2）. The pulse encoder on the motor detects shaft revolutions, and sends position signals to the CNC system（semi-closed loop control）. The conical ring couplings are used to

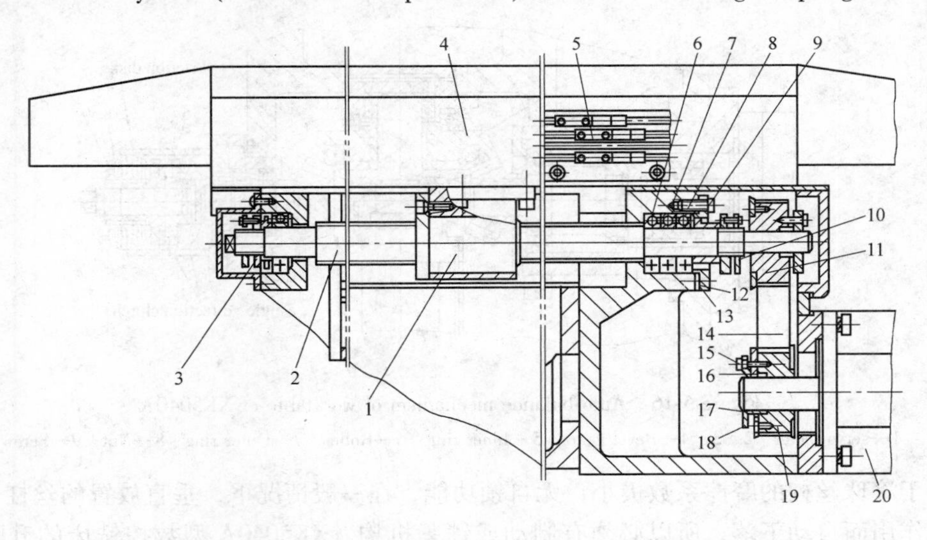

Figure 5-15　Longitudinal transmission mechanism of worktable

1, 3, 10—Nuts　2—Ball screw　4—Worktable　5—Stroke limiter　6、7、8—Bearings
9、15—Screws　11、19—Cog pulleys　12—Flange　13—Washer　14—Cog belt
16—Outer conical rings　17—Inner conical rings　18—End cover　20—AC servo motor

connect between the pulley hub and the motor shaft, and also between another pulley hub and ball screw for better transmission rigidity.

Fastening the set screw (9) produces preloading force on outer rings of bearings (6, 7, 8) through the flange plate (12). Preloading force is determined by the thickness of the washer (13).

Fastening the adjusting nut (3) applies a pre-tensile force onto the ball screw (2) to improve transmission rigidity and reduce thermal extension.

The stroke limiter (5) determines moving distance of the worktable.

如图 5-15 所示，交流伺服电动机 20 的轴上装有同步带轮 19，丝杠右端装有同步带轮 11，电动机通过同步带 14 驱动丝杠，从而使底部装有螺母 1 的工作台 4 移动。编码器检测伺服电动机 20 的转角并反馈至数控装置，形成半闭环控制。同步带轮与电动机轴及丝杠之间均采用锥环联轴器连接，以免出现反向间隙。

旋紧螺钉 9 时，法兰盘 12 压紧轴承外圈，对轴承产生预紧力，预紧力可通过修磨垫片 13 的厚度进行调整。调整丝杠左端的螺母 3 可使丝杠产生预拉伸，以提高丝杠的刚度和减小丝杠的热变形。

3. Worktable elevating and auto-balance mechanism

A brake or a lock mechanism is necessary for the vertical feeding transmission system since ball screw-nut pair cannot perform self-locking. Moreover, when the worktable is lifted, the servo motor must overcome the weight of the worktable thus a large output torque is required. When the worktable goes downwards, it drives the motor due to gravity. To even the motor torque output between upward and downward motion of the worktable, and to brake the worktable in case of power off, an auto-balance mechanism (see Figure 5-16) is used.

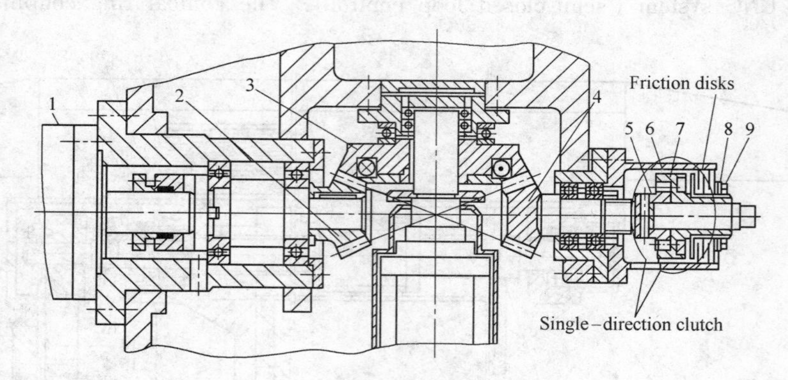

Figure 5-16　Auto-balance mechanism of worktable of XK5040A

1—servo motor　2、3、4—Bevel gears　5—Inner ring　6—Roller　7—Outer ring　8—Nut　9—Screw

由于滚珠丝杠的摩擦系数极小，无自锁功能，在一般情况下，垂直放置的丝杠会因部件的重力作用而自动下落，所以必须有制动或锁紧机构。XK5040A 型数控铣床的升降台自动平衡装置工作原理如下：

如图 5-16 所示，伺服电动机 1 通过锥齿轮 2 和 3 带动升降丝杠转动，实现工作台的上升或下降。锥齿轮 4 由锥齿轮 3 带动，并带动单向离合器的内环 5。当工作台上升时，内环的转向是使滚子 6 和外环 7 脱开的方向，外壳不转，摩擦片不起作用；当工作台下降时，内

环的转向是使滚子 6 楔在内环与外环 7 之间的方向，外环 7 随着锥齿轮 4 一起转动。经过花键与外壳连在一起的内摩擦片，与固定的外摩擦片之间产生相对运动，产生摩擦阻力，平衡重力的作用。摩擦阻力的大小可以通过螺母 8 来调整。

The inner ring of the single-direction clutch rotates with the shaft that connects the bevel gear (4) (see Figure 5-16), and the outer ring is fixed on another shaft that carries friction disks. When the inner ring rotates in clockwise, the rollers are compressed and produce friction between the inner and the outer rings to engage the clutch (see Figure 5-17a). In anticlockwise rotation, the rollers do not work and the clutch is disengaged (see Figure 5-17b).

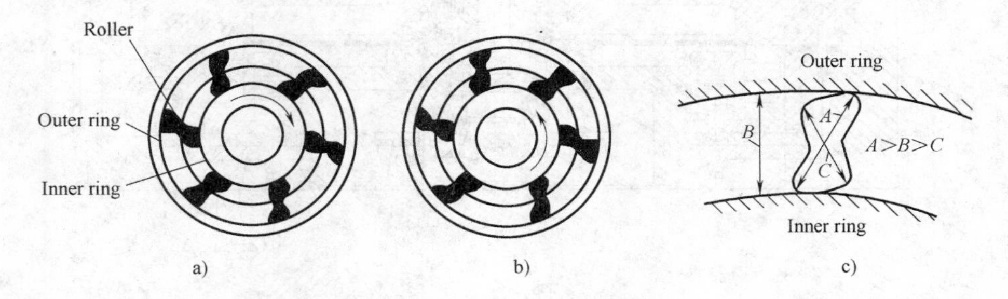

Figure 5-17 Work principle of single-direction clutch

The work principle of the auto-balance mechanism is shown in Figure 5-18.

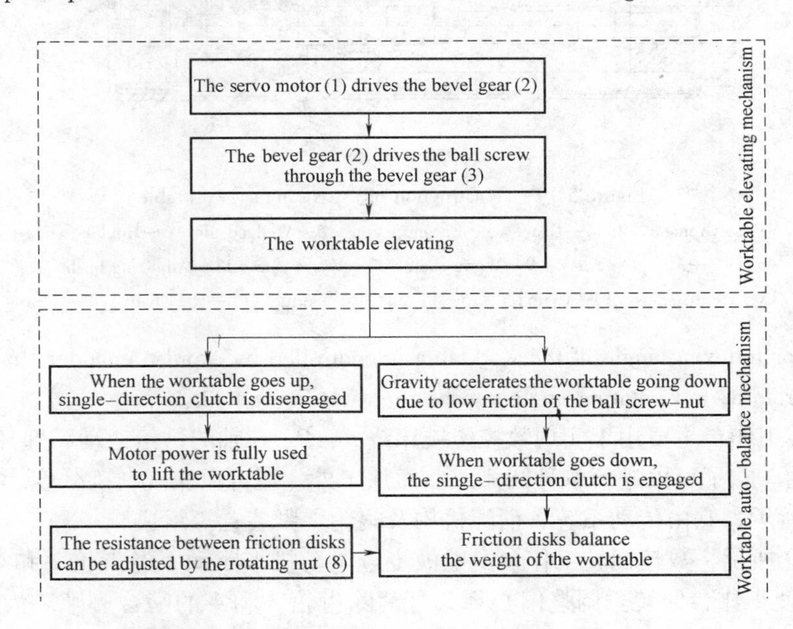

Figure 5-18 The work principle of the auto-balance mechanism

4. CNC rotary worktable

Circular feeding motion of *A*, *B*, *C*-axis is usually realized by a rotary worktable. Most boring-milling machines are equipped with CNC rotary worktables.

Besides feeding motion for contour machining, CNC rotary worktables can also perform accurate indexing. A typical rotary worktable consists of a transmission system, clearance eliminating mechanisms and wormgear clamps.

To improve transmission accuracy, the clearance eliminating mechanisms have been considered in the system. The eccentric ring (3) is used to adjust the relative position of the gears (2, 4) to eliminate the clearance between them. The worm shaft (9) has a continuous varying pitch. By axial adjustment, the clearance between the worm gear and the worm shaft can be minimized (see Figure 5-19).

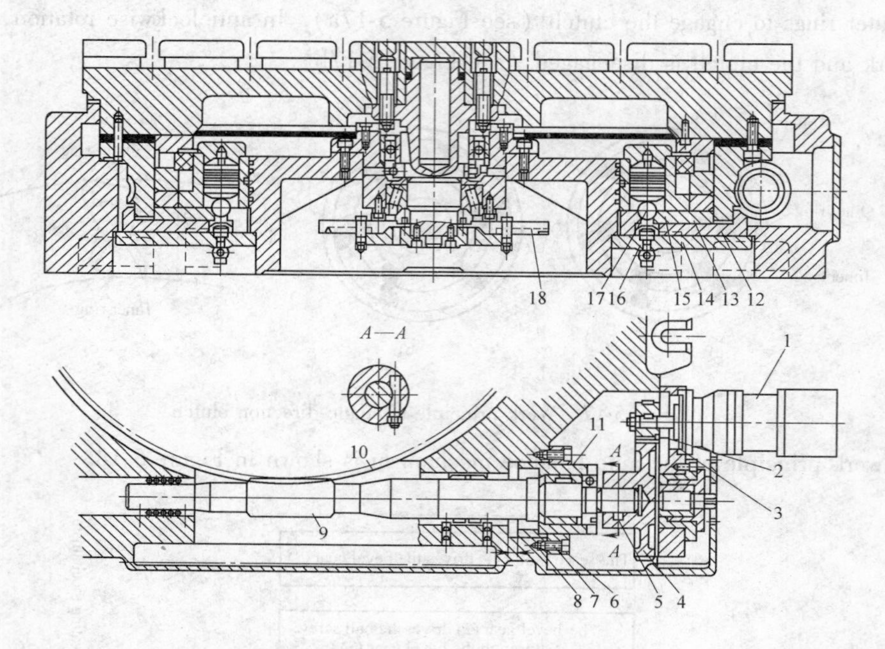

Figure 5-19 Construction of a CNC rotary worktable

1—servo motor 2, 4—Gears 3—Eccentric ring 5—Wedged pin 6—Holder (压块) 7—Nut 8—Screw 9—Worm shaft 10—Worm gear 11—Adjusting bush 12、13—Clamps 14—Cylinder 15—Piston 16—Spring 17—Steel ball 18—Grating

Rotating or indexing angle of the worktable is controlled by a pulse encoder (not shown in the picture). Therefore, any rotating angle can be precisely controlled.

数控回转工作台主要用于实现数控镗铣床绕 *A*、*B*、*C* 轴的进给运动，即环绕 *X*、*Y* 和 *Z* 轴的圆周运动。除了可以进行各种圆弧或曲面加工外，还可以实现精确分度。数控回转工作台如图 5-19 所示，它由传动系统、消隙机构及蜗轮夹紧装置等组成。

如图 5-19 所示，数控回转工作台由电液步进电动机 1 驱动，经齿轮 2 和 4 带动蜗杆 9，通过蜗轮 10 使工作台回转。调整偏心环 3 可消除齿轮 2 和 4 的啮合侧隙。齿轮 4 与蜗杆 9 靠楔形拉紧圆柱销 5（*A*—*A* 面）来连接，以消除轴与套的配合间隙。蜗杆 9 为螺距渐厚蜗杆，通过移动其轴向位置来调整间隙。调整时，松开螺母 7 的锁紧螺钉 8 使压块 6 与调整套 11 松开，转动调整套 11 带动蜗杆 9 作轴向移动；调整后，锁紧调整套 11 和楔形拉紧圆柱销 5 来消除间隙。

工作台静止时，底座上均布的 8 个夹紧液压缸 14 上腔通入压力油，活塞向下移动，通过钢球 17 撑开夹紧块 12 和 13，将蜗轮锁紧。当工作台需要回转时，数控系统发出指令，

夹紧液压缸14上腔的油流回油箱，钢球17在弹簧16的作用下向上抬起，夹紧块12和13松开蜗轮，允许蜗轮和回转工作台按照数控系统的指令做回转运动。

数控回转工作台可任意回转和分度，由光栅18进行读数控制，因此能够达到较高的分度精度。

The work principle of the CNC rotary worktable indexing is shown in Figure 5-20.

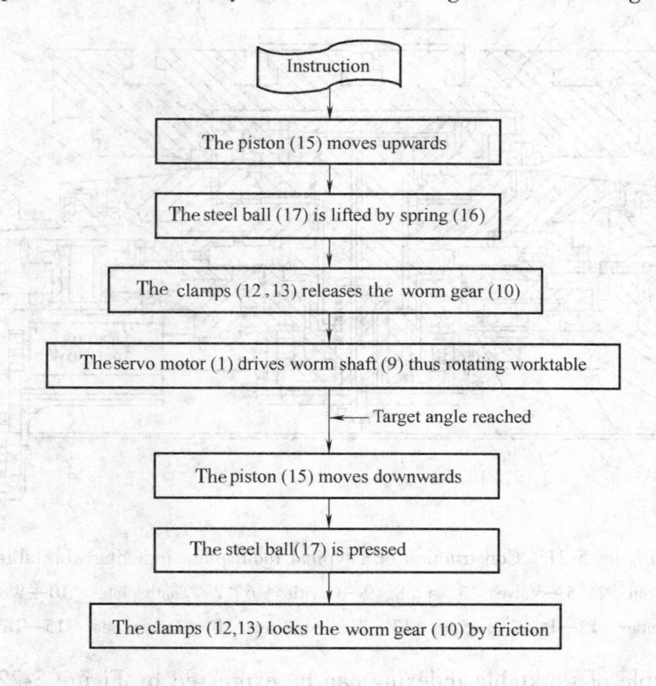

Figure 5-20 The work principle of the CNC rotary worktable indexing

5. Indexing worktable

Featured of simple construction and good accuracy, the tooth plate indexing worktable (Figure 5-21) is one of the most commonly used indexing worktable systems. A tooth plate indexing worktable system consists of a worktable, a lock cylinder and a pair of tooth plates, each of which usually has 120 or 180 triangular teeth. Besides indexing, the teeth on the plate realize circular and radial alignment of the plate pair.

The indexing process of the tooth plate indexing mechanism consists of preparation, worktable rotating, worktable positioning and clamping, and system reset.

The tooth plate indexing worktable has good accuracy of indexing and repositioning. However, high accuracy means high manufacturing cost. The indexing angle is restricted, which can be represented as:

$$\alpha = \frac{k360^\circ}{n}$$

Where, k is an integer, i. e. 1, 2, 3, ⋯, and n is the tooth number of the tooth plate.

如图5-21所示，鼠牙盘式分度工作台主要由工作台、夹紧液压缸及鼠牙盘等零件组成。其中，鼠牙盘是保证分度定位的关键零件，每个齿盘的端面均加工有相同数目的三角形齿，齿数一般为120个或180个，两齿盘啮合时能自动确定周向和径向的相对位置。一般有分度准备、分度动作和定位夹紧以及复位几个步骤。

鼠牙盘式分度工作台的工作特点如下：定位刚度好，重复定位和分度精度高，结构简单，但鼠牙盘制造精度高，且不能任意分度，只能分度可以除尽鼠牙盘齿数的角度。

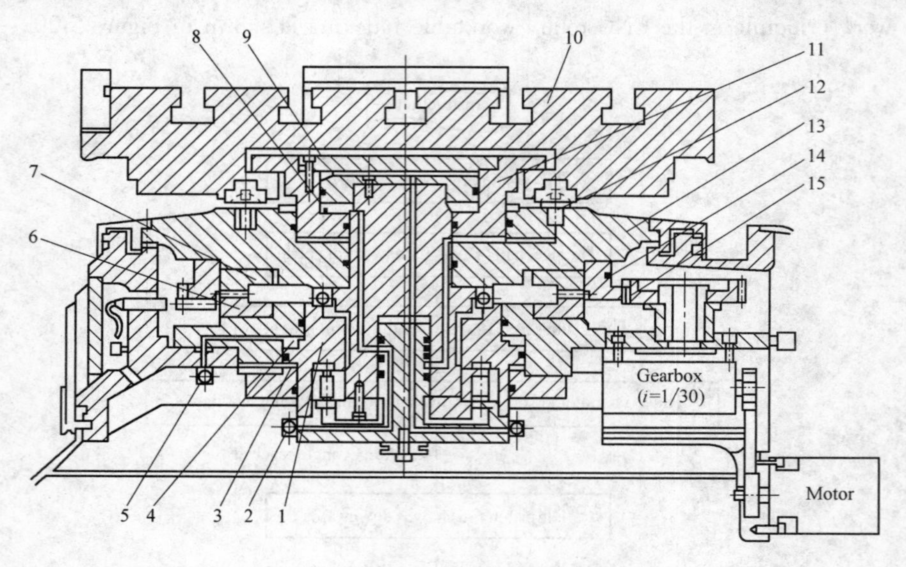

Figure 5-21　Construction of a typical tooth plate indexing worktable

1—Piston　2、5—Valves　3、4、8、9—Cavities　6、7—Tooth plates　10—Worktable
11—Cylinder　12—Positioning pin　13—Worktable seat　14—Driven gear　15—Driving gear

The work principle of worktable indexing can be expressed by Figure 5-22.

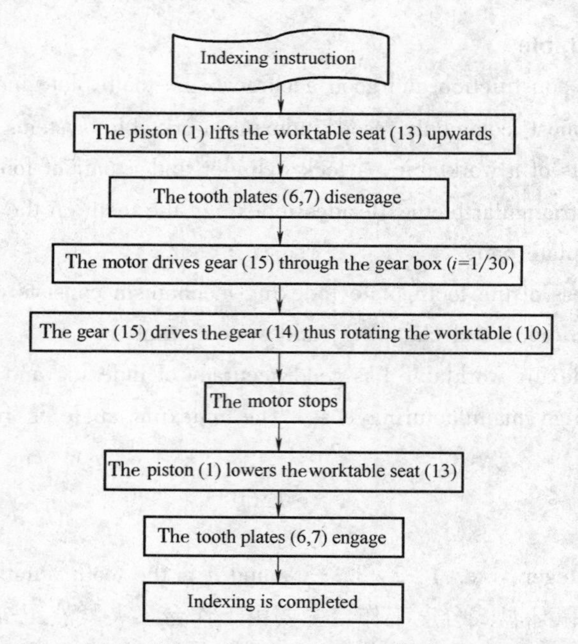

Figure 5-22　The work principle of worktable indexing

鼠牙盘式分度工作台分度过程：

当控制系统给出分度指令后，机床液压系统使压力油进入油腔3，使活塞1带动整个工作台台座13向上移动。台座13的上移使鼠牙盘6与7脱开。当控制系统给出转动指令时，电动机通过传动带和一个降速比为 $i=1/30$ 的减速箱带动驱动齿轮15和齿圈14转动，实现工作台座13的转动。当转过所需角度后，驱动电动机停止，压力油通过液压阀5进入油腔4，迫使活塞1向下移动并带动整个工作台台座13下移，使上下鼠牙盘啮合，进行精确定位，完成工作台的分度回转。

Many CNC milling machines apply exchangeable worktable to improve manufacturing efficiency. The principle of worktable exchange is expressed by Figure 5-23.

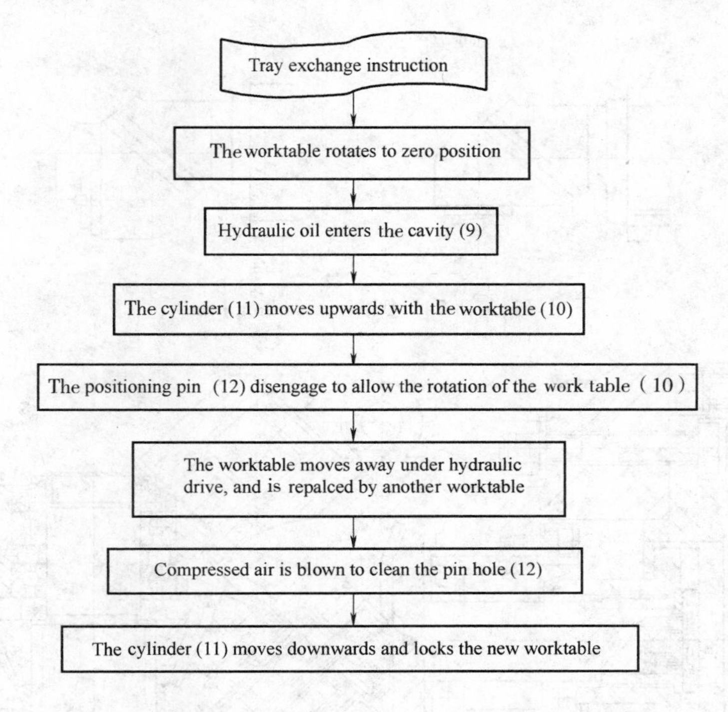

Figure 5-23 The principle of worktable exchange

工作台交换过程：

控制系统发出指令，使分度工作台返回零点，此时液压阀16接通，压力油进入油腔9，使得液压缸11向上移动，工作台脱离定位销12。当工作台被顶起后，液压系统带动工作台移出，与另一个待命的工作台交换位置。当新的托盘到达分度工作台上面时，空气阀接通，压缩空气经管路从工作台定位销12中间吹出，清除销孔中的杂物。同时电磁液压阀2接通，压力油进入油腔8，使液压缸11向下移动，并带动新的工作台夹紧在4个定位销12中，至此完成工作台交换。

6. Universal milling head (see Figure 5-24)

(1) Functions　Spindle direction can be adjusted by rotary surfaces *A* and *B* to allow the spindle axis to locate at any angle in hemisphere.

(2) Work principle　Figure 5-25 shows the typical construction of a universal milling head.

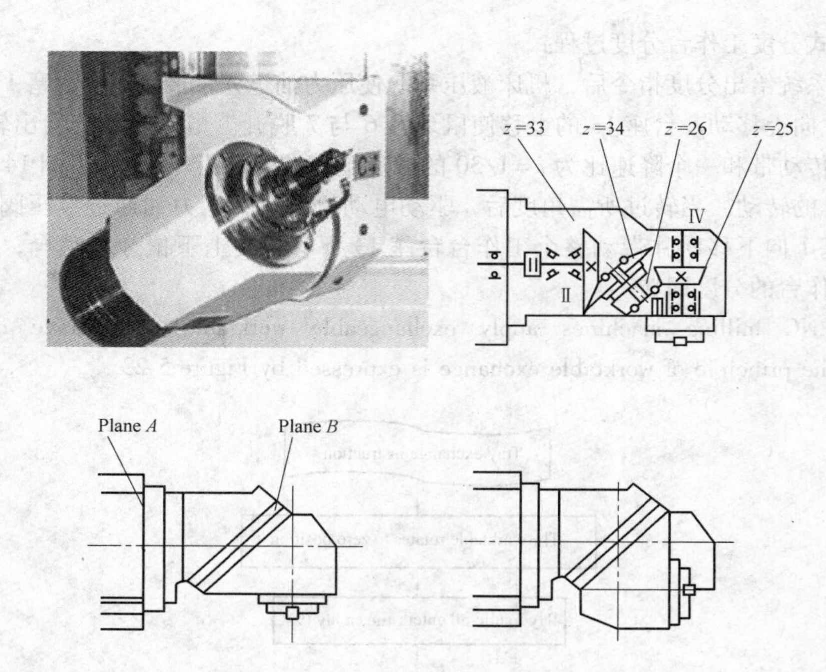

Figure 5-24 Universal milling head

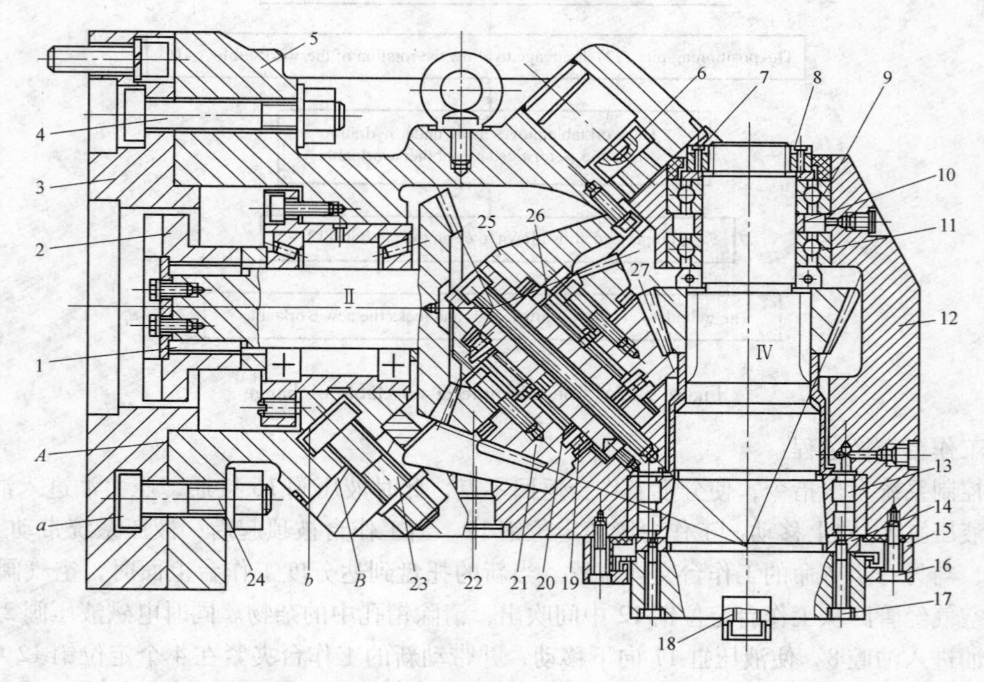

Figure 5-25 Typical construction of a universal milling head

1—Key 2—Connecting plate 3、15—Flange 4、6、23、24—T-bolt 5—Rear housing
7—Set screw 8—Nut 9、11—Angular contact bearings 10—Bush 12—Front housing
13—Roller bearing 14—Semicircular washer 16、17—Screw 18—Transverse key
19、25—Thrust bearings 20、26—Needle bearings 21、22、27—Bevel bears

The spindle direction can be manually adjusted.

After loosening the bolt (4) and the bolt (24), the milling head can be rotated about the axle Ⅱ, and at the same time, the bolt heads are performing circular moving in the T-groove (a) of the flange (3). The rotation takes place at the plane *A* (see Figure 5-20).

Milling head can rotate about the axle Ⅲ after loosening the bolt (6) and the bolt (23). Now the rotation takes place at the plane *B* (see Figure 5-24).

Synthetic rotations of the milling head at planes *A* and *B* allow the spindle axis to point at any angle in hemisphere.

(3) Bearing adjusting Purpose: to even distribution of load on rollers and to compensate bearing clearance due to wearing.

Adjusting procedure:

① Loosening the screw (16) to disassemble the flange (15).

② Loosening the screw (7) and the nut (8).

③ Pushing down the spindle Ⅳ for 2mm.

④ Disassembling half washer and grinding to a certain thickness to adjust preloading force for the bearing (13).

⑤ Preloading force for the bearings (9, 11) can be adjusted by the length of the bush (10).

⑥ Assembling.

Glossary

bevel gear		斜齿轮；锥齿轮
brake	[breik]	制动器
clamping	[klæmpiŋ]	（工具、刀具）夹紧
consideration	[ˌkənsidəˈreiʃən]	需要考虑的事
disassembling	[ˌdisəˈsembliŋ]	拆卸；分解
friction disk		摩擦片；摩擦盘
modeling	[ˈmɔdliŋ]	建模
profile	[ˈprəufail]	轮廓；外形
semicircular	[ˌsemiˈsəːkjulə]	半圆的
single-direction clutch		单向离合器
stroke limiter		行程限制器
synthetic	[sinˈθetik]	综合（性）的，合成的
tooth plate		齿盘
torsional	[ˈtɔːʃənəl]	扭转的
typical	[ˈtipikəl]	典型的
universal	[ˌjuːniˈvəːsəl]	万能的；通用的
worktable	[ˈwəːk，teibl]	工作台

Exercises

1. Calculate possible spindle rpm for XK5040A according to its transmission chain that is shown in Figure 5-13.

2. What is the main function of a universal milling machine? Is it used for machining workpieces that are in mass production?

3. How does the worktable of XK5040A realize auto-balance in vertical direction? If power supply is cut off during machining, what will happen to the worktable?

Chapter 6　Machining Center

6.1　Introduction of machining centers

6.1.1　Highlights of machining centers

① Modern machining centers are able to perform simultaneous control of at least three axes. They are suitable for 3D complex contouring machining.

② The automatic tool changer (ATC) which is a symbol of machining center significantly improves machining efficiency. It provides possibility of continuous machining of multi-procedure.

③ The CNC indexing rotary worktable can realize precise angular motion, e. g. 0. 1°. The rotary table with a horizontal spindle provides optimum perpendicular surface machining, most of which can be performed by clamping once. Moreover, with the rotation of the worktable, the machining surface gets closer to the spindle thus allowing a shorter length of the spindle to improve machining rigidity.

④ Intelligent CNC system provides software functions to simplify programming, e. g. tool compensation, machining circulation and instruction repeating. Some systems can even perform automatic programming.

⑤ Highly concentrated machining procedures significantly save time of workpiece clamping, measuring, transferring, storing, and machining tool adjusting. Therefore, machining center is 3-4 times efficient than conventional machine tools.

Continuous automatic machining also eliminates manual errors, thus is suitable for those workpieces that have complex contour, high accuracy requirement and frequent variety.

加工中心的基本特征：

① 具有至少三个轴的点位直线切削控制能力，特别适合复杂轮廓的切削。

② 具有刀具自动交换装置（ATC），这是加工中心机床的典型特征，是进行多工序加工的必要条件，能大大提高加工效率。

③ 具有高精度的分度回转工作台。这种回转工作台与卧式主轴相配合，使工件各垂直加工面与主轴最大程度地接近，主轴外伸少则刚性好。大多数加工中心机床都使用卧式主轴并配合回转工作台，以便一次装夹就能完成各垂直面的加工。

④ 具有各种保证加工过程自动化的辅助功能，还有如刀具补偿、固定加工循环、重复指令等功能，以简化程序编制工作。有些加工中心控制系统还能进行自动编程。

⑤ 工序高度集中。由于加工中心的上述特点，大大减少了工件的装夹、测量和机床的调整时间，减少了工件的周转、搬运和存放时间，使机床的切削效率比普通机床高 3 ~4 倍。

与普通机床相比，加工中心能排除工艺流程中的人为误差，特别适合加工形状复杂、精度要求高、品种更换频繁的工件。

6.1.2 Types of machining centers

1. Classifications by machining tasks

Machining centers can be classified into turning centers, drilling centers, grinding centers, milling/boring centers, and EDM centers. Usually, milling/boring centers are so-called machining centers, Shown in Figure 6-1.

按加工范围，加工中心可分为车削中心、钻削中心、镗铣中心、磨削中心和电火花加工中心等。一般镗铣加工中心简称加工中心。图 6-1 所示为五轴加工中心。

Figure 6-1 A five-axis milling center

2. Classifications by mechanical layouts

(1) Vertical machining center (VMC) Vertical machining centers indicate those machining centers with vertical spindle axis. They have relatively simple structure, less area occupation and economical price.

A VMC usually has a stationary column, and performs linear motions along X, Y, and Z-axis by a rectangular worktable. Equipped with a CNC rotary platform, helix machining is capable.

Figure 6-2 shows a vertical machining center.

立式加工中心是指主轴轴线为垂直设置的加工中心，其结构多为固定立柱式，工作台为长方形，具有 3 个直线运动坐标（沿 X、Y、Z 轴方向），无分度回转功能，适合加工盘类零件。如果在工作台上安装一个水平轴的数控回转台，就可用于加工螺旋线类零件。图 6-2 所示为一台立式加工中心的外观。

(2) Horizontal machining center (HMC) Horizontal machining centers indicate those machining centers with horizontal spindle axis. They are often equipped with indexing rotary worktables. A HMC usually has the controllability of 3-5 axes. The most common configuration is X, Y and Z linear axes plus a rotary axis (realized by worktable rotation). It is suitable for machining box structures since it can perform machining of all sides (except top and bottom) by clamping once.

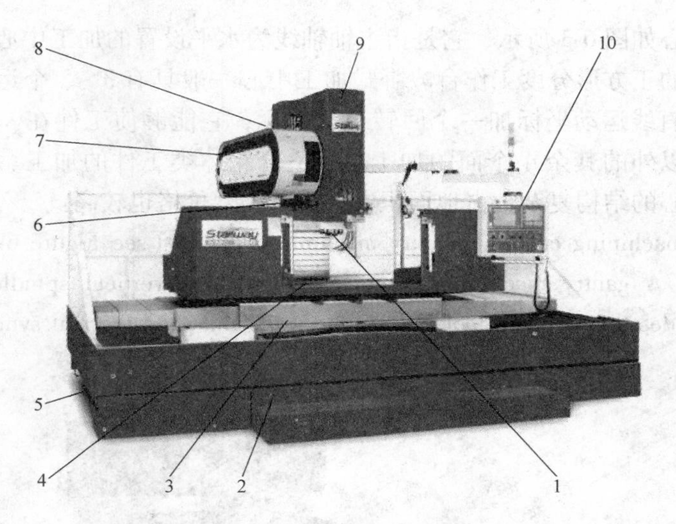

Figure 6-2 Vertical machining center

1—Spindle 2—Bed 3—Worktable seat 4—Worktable 5—Lubricant tank
6—Robot arm 7—Magazine 8—Column 9—Spindle case 10—Control panel

The relative motion between the cutting tool and the workpiece can be realized in different ways. For example, for an HMC with stationary column, the spindle case performs Y-axis motion and the worktable moves in X-Z plane. If the worktable is stationary, then the spindle case and the column move along X, Y, and Z-axis.

Compared to VMCs, HMCs have complex construction, larger area occupation. Moreover, they are much heavier and more expensive.

Figure 6-3 shows a horizontal milling center.

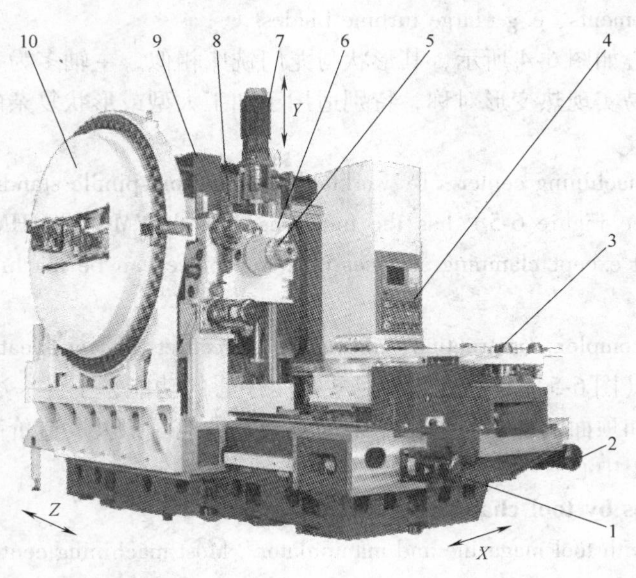

Figure 6-3 Horizontal milling center

1—Z-axis servo motor 2—X-axis servo motor 3—Worktable 4—Control panel
5—Spindle 6—Spindle case 7—Y-axis servo motor 8—ATC 9—Column 10—Magazine

卧式加工中心如图 6-3 所示，它是指主轴轴线为水平设置的加工中心，通常都带有可进行分度回转运动的正方形分度工作台。卧式加工中心一般具有 3～5 个运动坐标，常见的是 *X*、*Y*、*Z* 轴 3 个直线运动坐标加一个回转运动坐标，它能够使工件在一次装夹后就能完成除安装面和顶面以外的其余 4 个面的加工，最适合箱体类工件的加工。与立式加工中心相比，卧式加工中心的结构复杂，占地面积大，重量大，价格也较高。

（3）Gantry machining center　Gantry machining centers（see Figure 6-4）look like gantry milling machines. A gantry machining center usually has a vertical spindle. The twin column configuration provides the machine tool rigid frame work and good thermal symmetry.

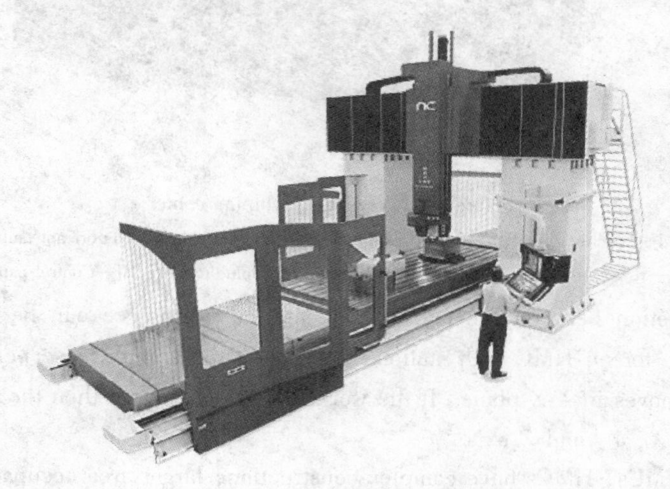

Figure 6-4　Gantry milling center

Gantry machining centers are suitable for machining large size complex parts, especially with high accuracy requirements, e. g. large turbine blades.

龙门式加工中心如图 6-4 所示，其形状与龙门铣床相似，主轴多为垂直设置。龙门型布局结构刚性好，容易实现热变形对称，特别适用于加工大型或形状复杂的精密工件，如汽轮机叶片等。

（4）Universal machining center　By worktable rotation or spindle standing/lying, a universal machining center（see Figure 6-5）has the functions of both VMC and HMC. With a universal machining center, all except clamping surfaces of a workpiece can be machined by clamping only once.

However, very complex constructions and high cost restrict their application.

万能加工中心（图 6-5）具有立式和卧式加工中心的功能，工件一次装夹就能完成除安装面外的所有侧面和顶面的加工。但由于结构复杂、占地面积大、造价高，它的应用范围远不如其他类型的加工中心。

3. Classifications by tool changing mechanism

（1）Equipped with tool magazine and manipulator　Most machining centers（e. g. JCS-018A）adopt this mechanism.

（2）Equipped with tool magazine but without manipulator　Magazine or part of magazine is

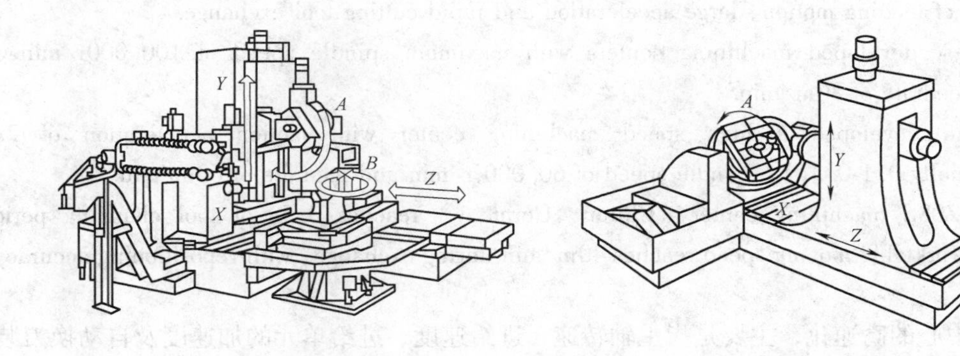

Figure 6-5 Universal milling centers

arranged where the spindle box can reach. Tool changing is realized by association of the magazine and the spindle box. The direction of a cutting tool is the same as that mounted on the spindle. Most small size machining centers (e. g. XH754 HMC) appliys this configuration.

(3) Turret magazine Most small size drilling centers (e. g. ZH5120 (see Figure 6-6)) adopt turret magazines for a simple construction.

根据换刀方式，加工中心可分为：

(1) 配置刀库、机械手的加工中心 其换刀装置由刀库和换刀机械手组成，加工中心普遍采用这种换刀方式。

(2) 无机械手加工中心 刀库一般布置在主轴箱可以运动到的位置，刀具存放在刀库中的位置方向必须与安装在主轴上时的方向一致，换刀时，由主轴移动到刀库，通过刀库和主轴箱的配合动作来完成存放刀具。

(3) 转塔刀库式加工中心 小型立式加工中心一般采用转塔刀库。这类加工中心主要以孔加工为主，如 ZH5120 型立式钻削中心，如图 6-6 所示。

Figure 6-6 ZH5120 drilling center

4. Classifications by CNC system characteristics

1) By motion axes controllability, e. g. three-axis machining center, five-axis machining center.

2) By servo mechanism, e. g. closed loop control, semi-closed loop control.

按数控系统的不同，加工中心有两种分类方法。一种是根据坐标轴的控制能力分类，如三坐标加工中心、五坐标加工中心；另一种是根据控制信号反馈方式，即半闭环加工中心和闭环加工中心。

6. 1. 3 Development tendency of machining center

High spindle rotation speed, high machining feedrate, high motion acceleration, rapid tool changing and worktable changing process are the development targets of modern machining centers.

(1) High speed Here the "high speed" means not only high spindle rotation speed, but also

high speed of feeding motion, large acceleration and rapid cutting tool exchange.

Japanese developed machining centers with maximum spindle speed of 100,000r/min and maximum feedrate of 80m/min.

German developed a high speed machining center with motion acceleration of 2.5g (conventionally 0.1-0.3g), spindle speed of 60,000 r/min and feedrate of 60m/min.

The FZ08S machining center (Chiron, Germany) reaches 0.5s of tool changing period. Recently, worktable moving speed reaches 40m/min during exchange, with repositioning accuracy of 3μm.

加工中心的高速化，主要是指主轴转速、进给速度、进给单元的加速度及自动换刀装置和自动托盘交换装置的高速化。

(2) Further increasing accuracy Swiss DIXI 1280TCA machining center reaches coordinate positioning accuracy of ±0.003mm/500mm in linear motion and 0.001° of rotation.

瑞士迪克西（DIXI）公司的DIXI 1280TCA型精密加工中心，其坐标定位精度已达到每500mm 行程 ±0.003mm，*B* 坐标（回转工作台）精度已达到3″。

(3) Improved intelligence Self-diagnosis function is involved in most modern machining centers. It is the symbol of intelligence improvement. Equipped with thermal sensors, acoustic sensors, and electric sensors, those machine tools have the functions of motion position correction, real time monitoring of cutting, standby cutting tool managing, temperature compensation, etc. For example, German WERNER TC series HMC can monitor spindle power, cutting load, tool length and tool damage (by sonar inspection) during machining process.

Advanced CNC device can meet the requirements of better accuracy and machining speed, and can also control peripheral equipments, e.g. robots and measuring instruments to improve machining efficiency.

加工中心功能的完善表现在越来越完善的自诊断功能。温度传感器、声传感器和电流传感器等使机床具有一定的人工智能，尽可能地减少加工中的故障。如德国 WERNER 公司的TC 系列卧式加工中心，它采用了主轴功率监控、切削负荷监控、刀具长度监控和声纳技术检测刀具破损情况等新技术，从而使加工中心的使用更加安全、可靠。

不断开发出的高精度、高速度、高效率的数控装置，把控制机器人、测量、上下料等功能都纳入到 CNC 内。

6.2 Automatic tool changer

6.2.1 Rotary turret

The turret carries many spindle heads, each of which is pre-installed with a cutting tool, e.g. drill, reamer, milling tool etc. Only the spindle head that has been rotated to working position connects transmission. Figure 6-7 shows a rotary turret. When it is mounted on a machining center, it looks like the one shown in Figure 6-8.

Figure 6-7 A rotary turret head

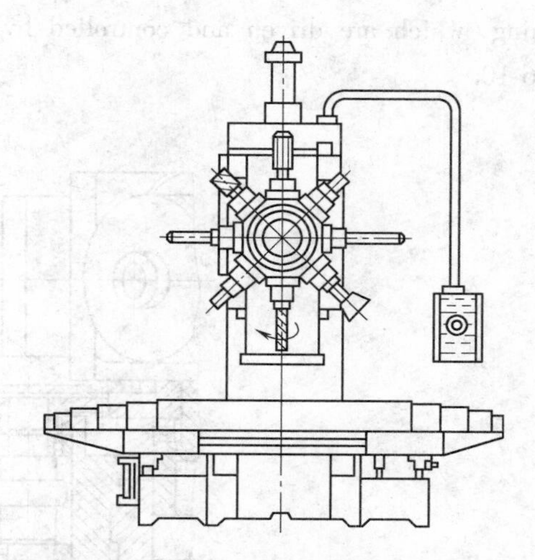

Figure 6-8 A machining center with automatic tool changer

(1) Construction of a turret head Figure 6-9 shows the construction of a horizontal turret head. Eight tool spindles that have the same construction are installed in the turret. The front bearing housing 2 and the spindle 1 have been assembled as an integral unit before installation for a convenient clearance adjusting. The pushing lever 12 can remove cutting tool through the action of the rod 14. The turret assembly is hydraulically pressed on the base. The transmission path between power source and cutting tool is: Motor-gearbox-gear 4-gear 13.

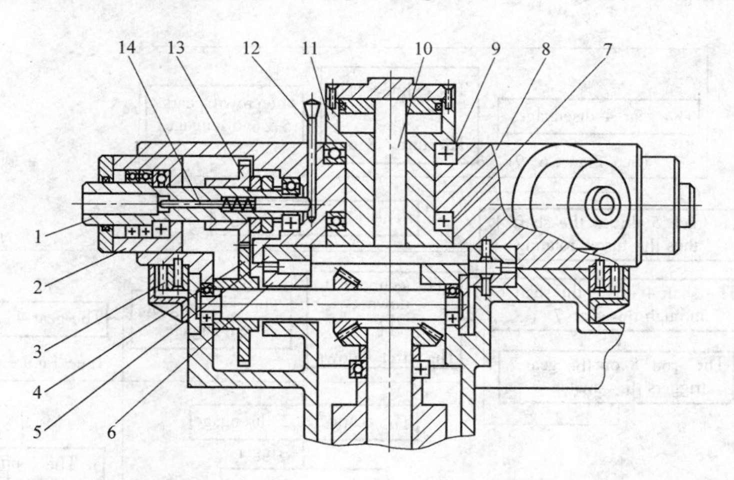

Figure 6-9 Construction of an eight-position turret

1—Spindle 2—Front bearing housing 3—Driving gear 4—Turret gear 5、6—Crown gears 7、9—Thrust ball bearings 8—Turret 10—Piston bar 11—Cylinder 12—Lever 13—Spindle gear 14—Rod

图 6-9 所示为卧式八轴转塔头结构。转塔头内均布八根刀具主轴，结构完全相同，前轴承座 2 连同主轴 1 作为一个组件整体装卸，便于调整主轴轴承的轴向和径向间隙。按压操纵杆 12，通过顶杆 14 可卸下主轴孔内的刀具。刀具主轴由电动机经变速机构、传动齿轮、滑移齿轮 4 到齿轮 13 进行传动。当压力油进入液压缸下腔时，转塔头即被压紧在底座上。

(2) Tool changing procedure of a turret head Tool changing is realized by turret rotating and

positioning, which are driven and controlled by a mechanical-electrical system that is shown in Figure 6-10.

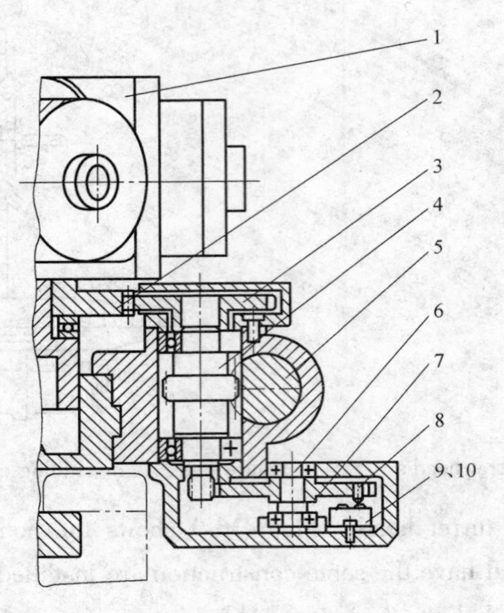

Figure 6-10 Transmission of turret head rotation

1—Turret 2—Turret gear 3—Driving gear 4—Driving shaft 5—Rack
6—Short Shaft 7—Gear 8—Rod 9、10—Positioning switches

Tool changing procedure can be illustrated by Figure 6-11.

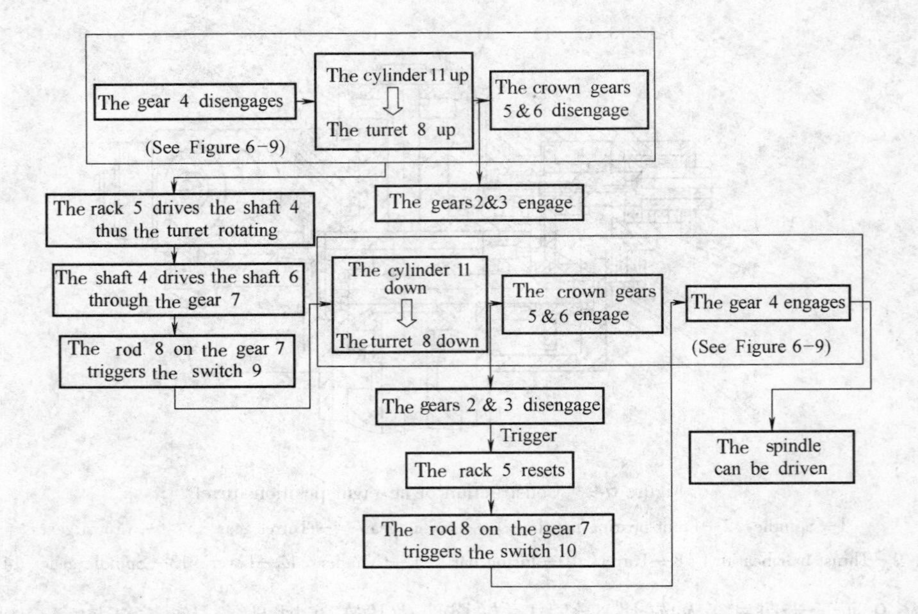

Figure 6-11 Tool changing procedure

6.2.2 Manipulator automatic tool changer

A manipulator tool changer consists of manipulator and magazine. They have been applied on

most modern machining centers, because they allow the tool magazines to have more flexible configurations and arrangements.

Manipulator can be in many forms, for example, single-arm, double-arm, a master arm with an assistant arm, etc. Tool changing process is rapid, usually 0. 5-2s.

目前大多数加工中心都配有机械手自动换刀装置。机械手自动换刀装置一般由机械手和刀库组成。其刀库的配置、位置及数量的选用要比无机械手的换刀装置灵活得多。可以根据不同的要求，配置不同形式的机械手，其换刀动作迅速，仅需0. 5 ~2s。

6. 2. 3 Magazine

1. Magazine types

Machining tasks of the machining center determine the configuration and capacity of the magazine.

The configuration of automatic tool changer depends on the relative position between the spindle and the magazine, and also the manipulator motion types. Figure 6-12 shows the configurations of some commonly used tool magazines.

由于刀库位置和机械手换刀动作的不同，自动换刀装置的结构形式也多种多样。刀库的形式和容量主要取决于机床的工艺范围。图6-12 所示为几种常见的刀库结构形式。

(1) Linear magazine (see Figure 6-12a) Tools are in linear arrangement. As the simplest magazine, it can only hold 8-12 tools. Linear magazines are usually applied on CNC lathe, and sometimes CNC driller.

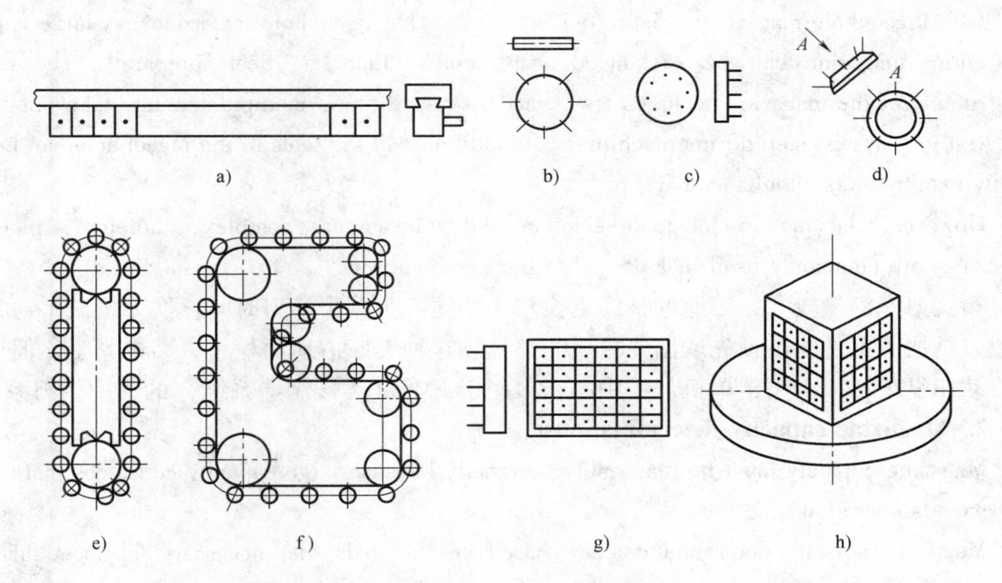

Figure 6-12 Configurations of commonly used magazines

a) Linear magazine b) Circular magazine with radial tool arrangement

c) Circular magazine with axial tool arrangement d) Circular magazine with conical tool arrangement

e) Chain magazine f) Extended chain magazine g) Pigeonhole magazine h) Multi-surface pigeonhole magazine

直线式刀库中，刀具呈直线排列，结构简单，但存放刀具数量有限（一般为 8 ~ 12 把），多用于数控车床，也可用于数控钻床。

(2) Circular magazine　A circular magazine can store up to 50-60 tools, and can be in many forms.

In Figure 6-12b, tools are in radial arrangement. The magazine usually locates on the top of the machine tool column since it occupies much space.

In Figure 6-12c, tools are in axial arrangement. The magazine usually locates next to spindle. Both vertical and horizontal placement are applied.

In Figure 6-12d, tools are in conical arrangement. It usually locates on the top of the column with the axis inclined.

圆盘式刀库如图 6-12b ~ d 所示，其存刀量可达则 50 ~ 60 把，并且有多种形式。刀具径向布置的圆盘刀库占空间较大，一般置于机床立柱上端。刀具轴向布置的圆盘刀库常置于主轴侧面，刀库轴心线可垂直放置，也可水平放置，其使用较为广泛。刀具为伞状布置的圆盘刀库，多斜放于立柱上端。

(3) Chain magazine (see Figure 6-12e, f)　Chain magazines are popular for their large and flexible capacity. Tool holders are fixed on chain pins. A single-chain magazine is capable to hold 30-60 tools. Using multi-chain or extended chain further increases magazine capacity.

链式刀库也是应用广泛的一种刀库形式，如图 6-12e、f 所示。链式刀库的刀座固定在链节上，常用的有单排链式刀库，一般存刀量为 30 ~ 60 把。若要进一步增加存刀量，则可使用多排链条或加长链条的链式刀库。

(4) Pigeonhole magazine (Figure 6-12g, h)　The pigeonhole magazine has large capacity. The entire magazine can be exchanged with another that has been prepared. The compact construction of the magazine occupies less space. To save time, manipulator may take out the tool and wait for the next step during machining. In addition, all the tools in the pigeonhole can be taken out by manipulators simultaneously.

However, the motions of tool selection and transfer are complex, therefore, pigeonhole magazines are commonly used in FMS rather than a single MC.

箱式刀库容量较大，刀库的整体更换较为便捷。为减少换刀时间，换刀机械手通常在前一把刀具加工工件时，预先取出要更换的刀具。这种刀库占地面积小，结构紧凑，但选刀和取刀动作复杂，很少用于单机加工中心，多用于 FMS（柔性制造系统）的集中供刀系统。

2. Magazine capacity determinations

Magazine capacity must be reasonably designed, because large capacity consumes more space, and extends operation time.

Magazine capacity determinations are based on the tools that necessary for most machining tasks. A survey shows that a capacity of 10-40 tools is economy, otherwise, cutting tool utilization rate drops while complexity rises (see Figure 6-13).

刀库的容量并不是越大越好，应当依照该机床大多数工件加工时需要的刀具数量来确定。据统计，80% 以上的工件完成其全部加工只需 40 把左右的刀具就足够了。为实用起见，刀库的容量一般为 10 ~ 40 把，盲目地加大刀库容量，会使刀库的利用率降低，结构过于复

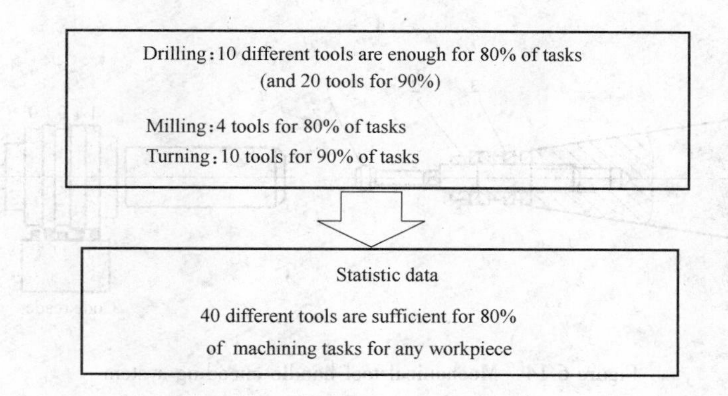

Figure 6-13 Cutting tool utilization rate

杂，并使选刀时间加长，从而造成浪费。

3. Cutting tool selection modes

(1) Sequential picking up All the tools that required are installed into tool holders in sequence before machining. The machine tool will pick up the tools one after another from the first to the last. Sequential picking up is easy to drive and control, but for different workpieces, tools need to be reinstalled.

Cutting tool installation is tedious manual work, and tool dimension errors result in unstable machining accuracy. Usually, sequential picking up mode is applied on economic CNC machine tools, and is suitable for mass production.

顺序选刀方式是将加工所需要用到的刀具按顺序装入刀套。加工时，机床顺次取用刀具。顺序选刀方式易于控制和驱动，但当改变工件时，需要根据新的工序重新装刀。装刀是一项烦琐的工作，且刀具的尺寸误差会影响加工精度的稳定性。顺序选刀方式通常用于经济型机床进行的大批量加工。

(2) Tool handles encoding Each tool has an unique code. The code identifier of the machine tool can recognize a certain tool when the tool code matches the instruction. The encoded cutting tools can be used whenever needed thus less magazine capacity is required. After using, the cutting tool is unnecessary to be placed into its original holders. Moreover, the mistake only occurs during programming thus better stability of accuracy.

刀柄编码方式是对每一把刀具的刀柄进行编码。机床的选刀识别装置会根据指令选取相应的刀具。与顺序选刀相比，刀柄编码的好处是每把刀具都可以重复使用，从而占用更少的刀库空间。使用后，刀具也不必放回原来的位置。采用刀柄编码方式，错误只可能发生在编程阶段，因此具有更高的精度稳定性。

Figure 6-14 shows the construction of an encoded tool handle. The code rings on tool handle have different radius. Small radius represents number “0” and large radius for “1”. The code is not acceptable if all the code rings are in small diameter, since it is confusing with “nothing”. Therefore, n rings can compose a number of codes ($2^n - 1$).

The code rings require a mechanical code reader that has simple construction. Since physical contact is unavoidable during working, a mechanical code reader has limited reading speed and low reliability.

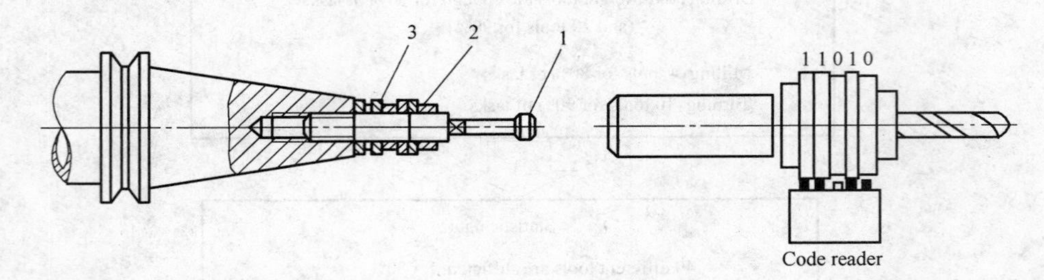

Figure 6-14 Mechanical tool handle encoding system

1—Pulling shaft 2—Lock nut 3—Code rings

刀柄编码是通过一组半径不同的码环组合实现的（图 6-14），小环代表“0”，大环代表“1”。为了避免识别装置误判“无刀具”，编码不能为全部为 0。因此，n 个码环可以组成 (2^n-1) 个编码。

机械式编码和读码器结构简单，由于其工作过程采用物理式接触，因此读取速度和可靠性受到了限制。

The magnetic encoding system（see Figure 6-15）eliminates the physical contact during code reading, therefore, it is silent and reliable. the magnetic code reader has no mechanical contact and abrasion during working, therefore, it has a much longer service life. The principle of encoding is similar, i. e. code rings are made of magnetizer and non-magnetizer to represent “0” and “1”.

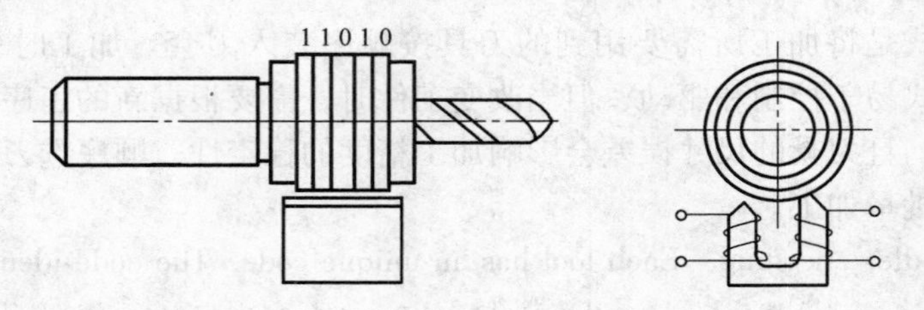

Figure 6-15 Magnetic tool handle encoding system

磁性编码读码装置（图 6-15）的编码原理与机械式相同，即码环用导磁与非导磁两种材料制造，分别代表“0”和“1”。磁性读码器在工作过程中不存在机械接触，因此噪声小，可靠性高，且无磨损，使用寿命长。

The code can also be realized by photoelectric technology, which is similar to bar codes that printed on every package in a supermarket. Besides the advantages of a magnetic encoding system, a photoelectric code reader（see Figure 6-16）has a much quicker response.

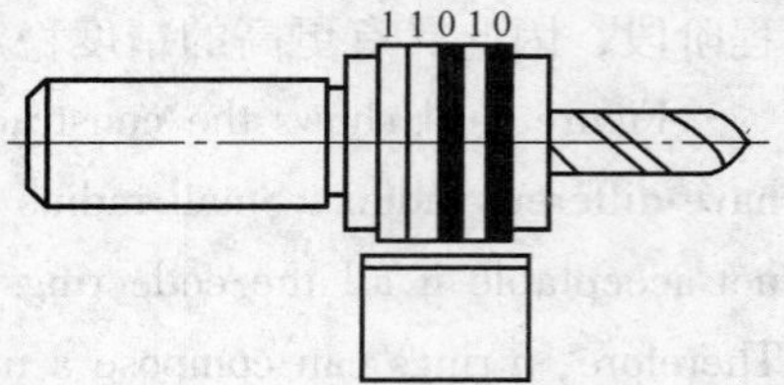

Figure 6-16 Photoelectric tool handle encoding system

编码读码还可以通过光电技术来实现，其特点是无噪声，可靠性高，响应速度快，使用寿命长。

(3) Tool holders encoding The principle of tool holders encoding is the same as that of tool handles encoding. This method simplifies the tool handle structure. The tool holders can

be in permanent encoding (see Figure 6-17) or temporary encoding.

For permanent encoding, each tool holder obtains its unique identifier. When a tool change signal comes, the magazine rotates until the code of certain tool holder matches the instruction. Since the system is looking for the tool holder rather than the cutting tool, tools must be mounted into corresponding holders, otherwise accident may occur.

刀套编码有利于简化刀柄的结构，其原理类似于刀柄编码。刀套编码分为永久编码（图 6-17）和临时编码两类。

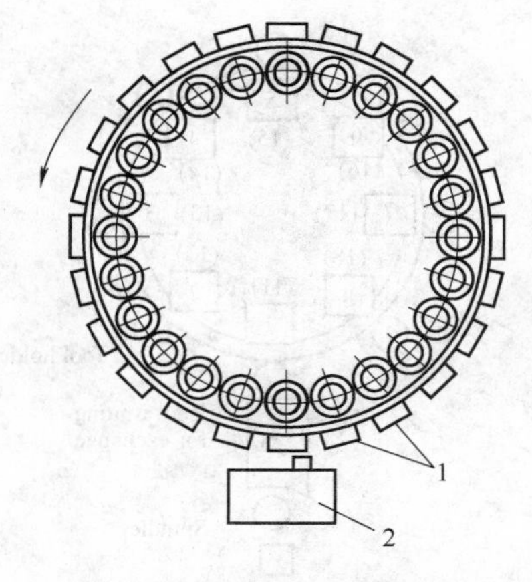

Figure 6-17 Permanently encoded tool holders

1—Code unit of tool handles 2—Code reader

永久编码是每个刀套均获得一个唯一且一般不作更改的识别码。选刀时刀库旋转，直到某一刀套上的代码与指令一致。由于读码器识别的是刀套，因此加工前刀具必须装入相应的刀套中，否则加工过程可能造成事故。

Temporary tool holder encoding eases the labor of tool mounting. Each cutting tool matches a key carrying a unique code (see Figure 6-18), and they must be used in pairs. When mount a tool into a holder, the corresponding key must be inserted into the key hole to encode the tool holder.

临时编码可减少人工装刀的工作量。其原理是每一把刀具都配有一把钥匙（图 6-18），将刀具装入刀套的同时，将钥匙插入刀套上的钥匙孔即完成了对刀套的编码。

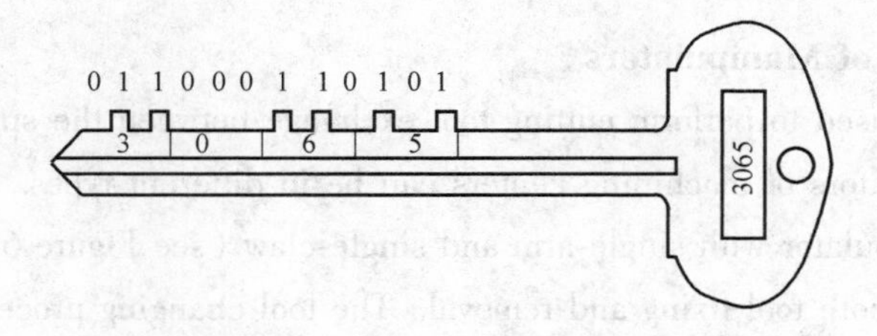

Figure 6-18 A key for temporary encoded tool holders

(4) Random picking up of cutting tools Figure 6-19 shows the principle of random picking up of cutting tools. Each location that may carry a cutting tool has an address, and each cutting tool has a digital code as the content in the address. The information is saved in as a datasheet in CNC device. Therefore, the datasheet actually is a reflection of tools' location (including the magazine and the spindle) in the machine tool. When a cutting tool is transferred, the content in certain address in the datasheet is transferred into another address with the same step. Therefore, each tool can be traced by the datasheet.

Once a datasheet has been used, any modification should be avoided.

For example, in Figure 6-19, address "TAB 0" represents the spindle hole, and "TAB 1" - "TAB 8" respectively represents eight tool holders of the magazine. The contents in the addresses correspond with tool numbers.

随机选刀是通过计算机软件来实现的。每一个刀位（包括刀套和主轴孔）均对应数据

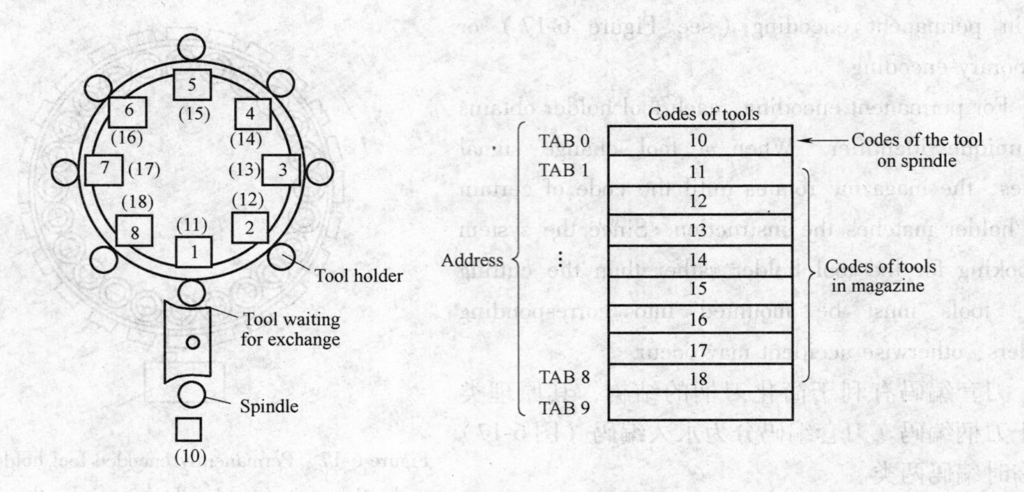

Figure 6-19　Software management of cutting tools

库中的一个地址，该地址中存储的内容即对应某一把刀具。因此，该数据库实际为刀具所在位置的映像。当刀具位置发生转移，数据库地址中存储的内容也同步更新。这样，每一把刀具的位置始终都由软件追踪管理。

如图 6-19 所示，地址“TAB 0”为主轴孔，“TAB 1” ~ “TAB 8”分别代表 8 个刀套。当刀具位置改变时，地址“TAB 0” ~ “TAB 8”中的刀号也会同步更新。

6.2.4　Manipulators

1. Classfications of Manipulators

Manipulators are used to perform cutting tool exchange between the spindle and the magazine. Tool changing manipulators of machining centers can be in different types.

(1) Rotary manipulator with single-arm and single-claw (see Figure 6-20a)　It has only one claw, which performs both tool fixing and removal. The tool changing procedure is time consuming.

单臂单爪回转式机械手（图 6-20a）可以转过不同的角度实现自动换刀，机械手的手臂上只有一个夹爪，刀库或主轴上的装刀及卸刀均由这一个夹爪来完成，换刀时间较长。

(2) Swing manipulator with single-arm and double-claw (see Figure 6-20b)　Two claws perform different actions: one for tool fixing and the other for removal. Compared to type (a), it saves 50% of operation time of tool exchange.

单臂双爪摆动式机械手（图 6-20b）有两个夹爪，一个夹爪只执行从主轴上取下“旧刀”送回刀库的任务，另一个夹爪则执行由刀库取出“新刀”送到主轴的任务，其换刀时间约为单臂单爪回转式机械手的一半。

(3) Rotary manipulator with single-arm and double-claw (see Figure 6-20c)　Claws locate at both ends of the arm. The cutting tools in the spindle and in the magazine would be gripped by the claw at the same time, and then the claw shaft rotates 180°for tool changing. The procedure is faster than both types (a) and (b).

单臂双爪回转式机械手（图 6-20c）的手臂两端各有一个夹爪，可同时抓取刀库及主轴

上的刀具，回转180°后又同时将刀具放回刀库及装入主轴。换刀时间比以上两种单臂机械手均短，是最常用的一种形式。

(4) Rotary manipulators pair (see Figure 6-20d) It can be regarded as a pair of single arm manipulators of type (a), one of which is to remove the cutting tool from the spindle and return to the magazine, and the other for mounting the tool on the spindle.

双机械手（图6-20d）相当于两个单臂单爪机械手，其中一个机械手从主轴上取下"旧刀"送回刀库，另一个则由刀库中取出"新刀"装入机床主轴，相互配合动作。

(5) Reciprocating manipulators pair (see Figure 6-20e) Each arm can extend and withdraw. The arms with the base would rotate. The manipulators shear a base, which can move along a way or rotate about a shaft. The flexible design simplifies the configuration between the magazine and the spindle.

双臂往复交叉式机械手（图6-20e）的两臂可以往复运动，并交叉成一定的角度。一个手臂负责卸刀，另一个负责装刀。双臂共用一个底座，底座可沿某导轨直线移动或绕某个转轴回转，以实现刀库与主轴间的换刀运动。

(6) Rotary manipulator with double-arm and end-claws (see Figure 6-20f) The manipulator catches the ends of cutting tools rather than cylindrical surfaces.

双臂端面夹紧式机械手（图6-20f）的夹紧位置是刀柄的两个端面，而不同于前几种机械手靠夹紧刀柄的外圆表面抓取刀具。

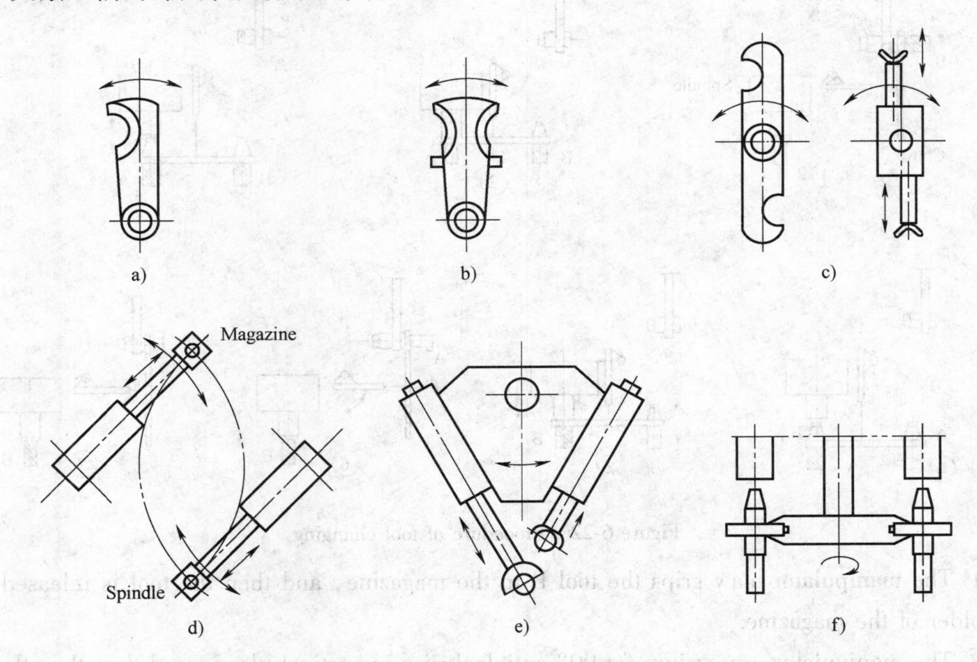

Figure 6-20 Commonly used manipulators

2. Manipulator motions

Most motions (e. g. tool removal, fixing, and arm rotation) of a rotary manipulator with double-arm and end-claws is driven by hydraulic power. The distance of motion and the angle of rotation are controlled by stroke limiters. Figure 6-21 shows the construction of the manipulator of an SOLON3-1.

双臂端面夹紧式机械手的拔刀、插刀动作，手臂回转大都由液压机构驱动。机械手的动作行程和回转角度可由行程开关来控制。图 6-21 所示为 SOLON3-1 卧式加工中心的机械手。

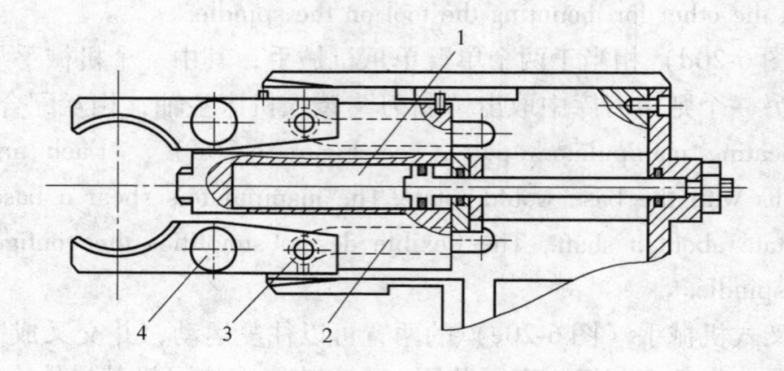

Figure 6-21 Construction of a manipulator of SOLON3-1 (HMC)

1—Cylinder 2—Guiding slots 3—Pins 4—Pivot

Figure 6-22 shows the seven steps of tool changing by a rotary manipulator with double-arm and end-claws.

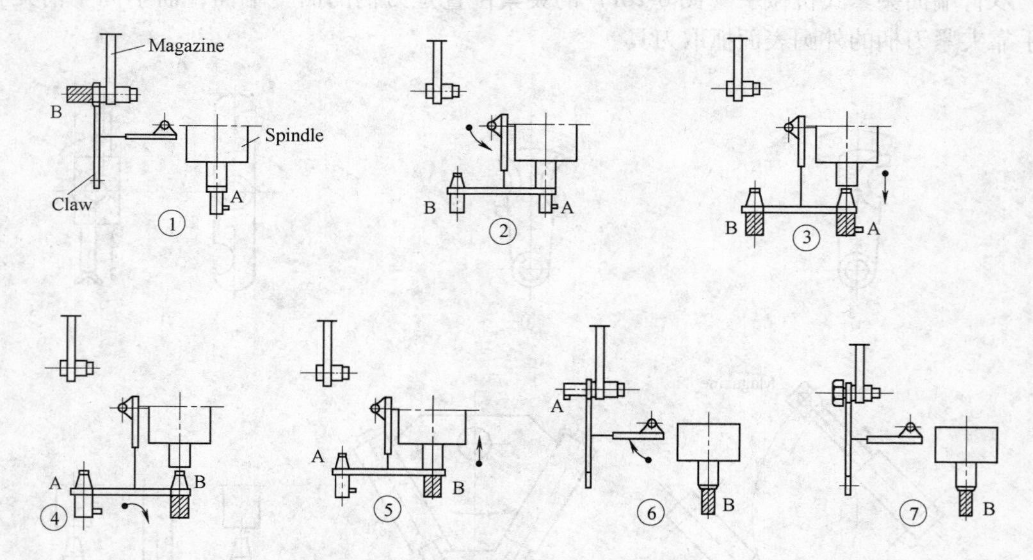

Figure 6-22 Procedure of tool changing

① The manipulator claw grips the tool B in the magazine, and then the tool is released by its tool holder of the magazine.

② The manipulator arm swings for 90° anticlockwise in vertical plane, and then the other claw grips the tool A in spindle.

③ The arm moves downwards to remove the tool A in the spindle.

④ The arm rotates of 180° in horizontal plane.

⑤ The arm moves upwards to install the tool B into spindle.

⑥ The claw releases the tool B. The arm then swings for 90° clockwise in vertical plane to return the tool A into magazine; magazine locks the tool A.

⑦ The claw withdraws and releases the tool A, and stops for the next tool to be selected.

其换刀过程的分解动作如图 6-22 所示。

① 夹爪伸出，抓住刀库上的刀具。刀库刀座上的锁板拉开。

② 机械手带着刀库上的刀具绕竖直轴逆时针方向摆动 90°，另一个夹爪伸出抓住主轴上的刀具。

③ 机械手前移，将刀具从主轴上取下。

④ 机械手绕自身水平轴转动 180°，将两把刀具交换位置。

⑤ 机械手后退，将新刀具装入主轴。

⑥ 夹爪回缩，松开主轴上的刀具。机械手绕竖直轴回摆 90°，将刀具放回刀库，刀库刀座上的锁板合上。

⑦ 夹爪缩回，松开刀库上的刀具，恢复到原始位置。

For a hydraulic driven tool changing system, sealing requirement conflicts agility of motion. Pneumatic driving causes large noise though it meets the requirement of fast motion. Moreover, both hydraulic and pneumatic systems need electromagnetic valves, which consume time for delay constant. Recently, mechanical tool changing driving mechanisms are rapidly developed.

目前，换刀机械手主要由液压或气动装置驱动。液压系统的密封性与动作灵敏性是相互矛盾的。气动系统动作敏捷，但工作噪声大。此外，液压或气动装置的控制元件电磁阀的延时常数也增加了时间消耗。近来，机械式换刀驱动系统得到了迅速的发展。

6.3 An introduction of JCS-018A VMC

6.3.1 Functions, features and parameters of JCS-018A VMC

1. Functions

The JCS-018A VMC can automatically and continuously perform machining operations such as milling, drilling, reaming, boring, expanding, tapping, etc. It is suitable for machining small size parts (e. g. plates, shells and moulds) in medium and small quantity. The JCS-018A VMC does not need many jigs, measuring instruments which are necessary on conventional machine tools thus shortening preparation period. Moreover, machining quality is significantly improved due to manual error elimination thus better manufacturing efficiency.

在 JCS-018A 型加工中心上，工件一次装夹后可以自动连续地完成镗、铣、钻、铰、扩、锪和攻螺纹等多种工序的加工，特别适合于小型板、盘、壳体和模具类零件的多品种中小批量加工。它可以节省在普通机床上加工所需的大量的工艺装备，缩短了生产准备周期；另一方面能够确保工件的加工质量，提高生产率。

2. Highlights

(1) Great cutting force The spindle of the JCS-018A is driven by a FANUC AC 12 motor through the cog belt for quiet transmission, reduced vibration and less thermal distortion. The spindle

has a wide constant power range and a large torque at low rotation speed. The rigid main framework of the machining center can withstand great cutting force.

JCS-018A 的主轴采用 FANUC AC12 型交流电动机，经一对同步带轮带到主轴，无齿轮传动，所以主轴运转时噪声低、振动小、热变形小。主轴转速的恒功率范围宽，低转速的转矩大。机床主体构件刚度高，可进行强力切削。

（2）High speed positioning　Through conical rings couplings and ball screw-nut system, the DC servo motors provide 14m/min feeding speed in *X* and *Y*-direction, and 10m/min in *Z*-direction. The plastic-coated guideways reduce vibration at high speed motion, and eliminate creeping at low speed. Therefore, the worktable has high accuracy and good stability of motion.

进给直流伺服电动机的运动经联轴器和滚珠丝杠副，使 *X* 轴和 *Y* 轴获得 14m/min、*Z* 轴获得 10m/min 的快速移动。贴塑导轨使机床在高速移动时振动小，低速移动时无“爬行”现象，具有很高的精度和良好的稳定性。

（3）Random tool selection　The rotary magazine is driven by a DC servo motor through wormgear reduction. Manipulator rotations, tool picking and installing motions are driven by hydraulic systems. Cutting tool management software traces the locations of all the tools. Before the magazine rotates, the system will make a judgment to ensure the rotation angle < 180° for time saving.

刀库回转运动由直流伺服电动机经蜗杆副驱动。机械手的回转、取刀和装刀机构均由液压系统驱动。刀具管理软件用于追踪所有刀具的位置，并由此作出判断，确保每次选刀时，刀库正转或反转角均不超过 180°。

（4）Integrated design of mechanical-electrical system　Both the control cabinet and the lubrication system are located on the bed and the column thus reducing area occupation and simplifying transporting of the machine tool.

机床的控制柜和润滑装置都安装在立柱和床身上，减少了占地面积，同时也简化了搬运和安装过程。

（5）Computer control system　The JCS-018A is controlled by a software-encapsulated CNC system. The system has small volume, low failure rate, high reliability and user-friendly interface. Both machining signals and the control system are under self-diagnosis, which provides users explicit monitoring and convenient inspection.

JCS-018A 采用软件型数控系统，体积小，故障率低，可靠性高，操作简便。机床信号系统和控制系统均有自诊断功能，便于用户监控和检查。

6.3.2　Transmissions of JCS-018A

The JCS-018A has five transmission chains for primary motion, feeding motions in *X*-, *Y*-, and *Z*-axis, magazine rotation. The arrangement of the transmission systems is shown in Figure 6-23。

JCS-018A 型加工中心的传动系统由五条传动链组成，即：主运动传动链，纵向、横向、垂直方向传动链以及刀库的旋转运动传动链，如图 6-23 所示。

1. Spindle transmission

The spindle rotation speed can perform continuous varying between 22.5-2250r/min.

The FANUC AC12 AC servo motor has maximum power of 15kW and rated power of 11kW. With power limiter, the spindle can provide maximum power of 7.5kW and rated power of 5.5kW.

The motor has a wide speed range of 45-4500r/min. Constant power output can be realized between 750r/min and 4,500r/min. After a belt reduction of 1:2, the spindle gets a constant power zone in 375-2,250r/min (see Figure 6-24a). 45-750r/min is a constant torque zone of the motor (and 22.5-375r/min for the spindle correspondingly). In this rotation speed range, the motor provides a continuous torque output of 140Nm (see Figure 6-24b).

主轴在22.5～2250r/min的转速范围内可以实现无级调速。FANUC AC12型交流伺服电动机在30min超载时的最大输出功率为15kW，连续运转时的最大输出功率为11kW。安装功率限制器后，电动机的额定输出功率为7.5kW（30min 超载）和5.5kW（连续运转）。

电动机转速范围为45～4500r/min，其中在750～4500r/min转速范围内为恒功率区域。经过同步带减速，主轴的恒功率转速范围为375～2250r/min。（图6-24a）。电动机转速在45～750r/min范围内为恒转矩区域，其连续运转的最大输出转矩为140 N·m，电动机最大输出转矩为191N·m（30min 超载）。主轴恒转矩区域的转速范围为22.5～375r/min。（图6-24b）

2. Feeding transmission

Feeding motions in X, Y, Z-direction of the JCS-018A are driven by FANUCBESK DC servo motors (1.4 kW each) through ball screw-nut transmissions. Any two axes can be controlled simultaneously.

Z-axis feeding is realized by spindle case

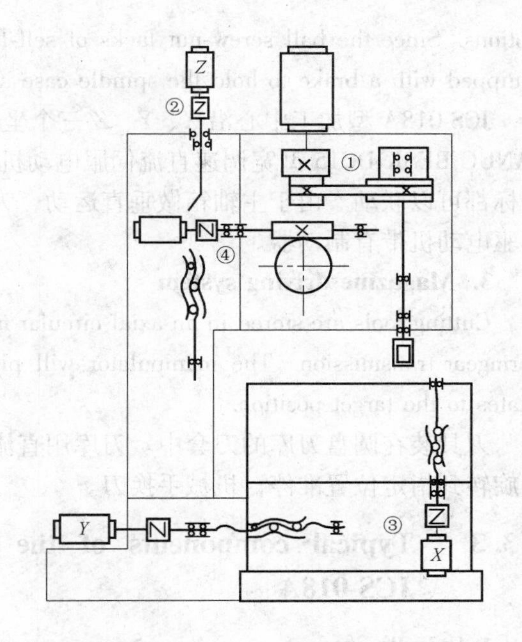

Figure 6-23 Transmission chains of JCS-018A

① Spindle transmission ② Spindle case elevating ③ Worktable moving in X-Y plane ④ Tool magazine rotation

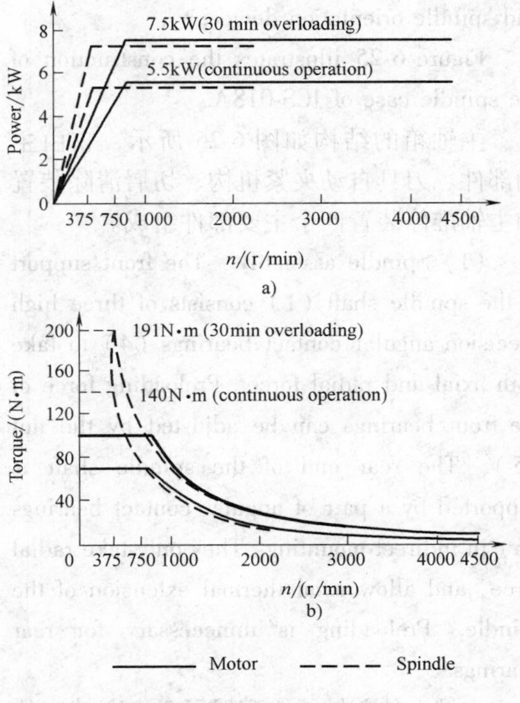

Figure 6-24 Spindle motor performance curves of JCS—018A VMC

a) Power curve b) Torque curve

motions. Since the ball screw-nut lacks of self-locking ability, the servo motor for *Z*-axis drive is equipped with a brake to hold the spindle case against gravity.

JCS-018A 型加工中心沿 *X*、*Y*、*Z* 三个坐标轴的进给运动分别是由三台功率为 1.4kW 的 FANUC BESK DC15 型宽调速直流伺服电动机直接带动滚珠丝杠旋转来实现的。其任意两个坐标都可以联动。由于主轴箱做垂直运动，为防止滚珠丝杠因不能自锁而使主轴箱下滑，*Z* 轴驱电动机带有制动器。

3. Magazine driving system

Cutting tools are stored in an axial circular magazine that is driven by a DC servo motor through wormgear transmission. The manipulator will pick up a required cutting tool when the magazine rotates to the target position.

刀具装在圆盘刀库的刀套中。刀库用直流伺服电动机经蜗杆副驱动，当需要换刀时，刀库旋转到指定位置准停，机械手换刀。

6.3.3 Typical components of the JCS-018A

1. Spindle case

The spindle case of the JCS-018A consists of four main parts, namely, spindle assembly, tool clamping device, chip cleaning blowpipe, and spindle orientation device.

Figure 6-25 illustrates the construction of the spindle case of JCS-018A.

主轴箱的结构如图 6-25 所示，它由主轴部件、刀具自动夹紧机构、切屑清除装置和主轴准停装置四个主要部件组成。

(1) Spindle assembly　The front support of the spindle shaft (1) consists of three high precision angular contact bearings (4) to take both axial and radial force. Preloading force of the front bearings can be adjusted by the nut (5). The rear end of the spindle shaft is supported by a pair of angular contact bearings (6) in indirect mounting. They only take radial force, and allow axial thermal extension of the spindle. Preloading is unnecessary for rear bearings.

主轴 1 的前轴承 4 配置了 3 个高精度的角接触球轴承，以承受径向载荷和轴向载荷。前轴承按预加载荷计算的预紧量由预紧

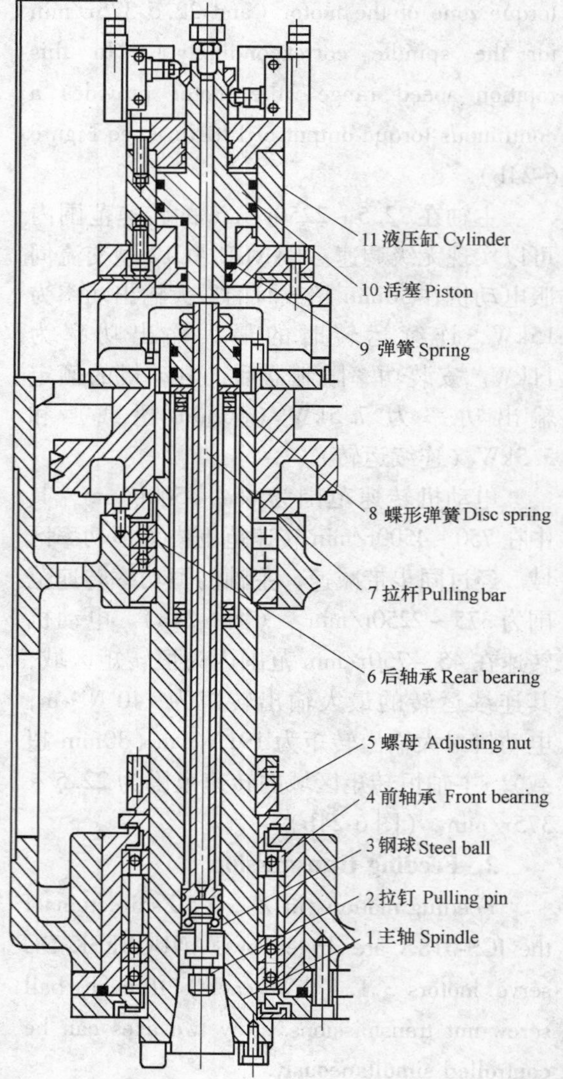

Figure 6-25　Spindle case construction of JCS-018A

螺母5来调整。后轴承6为一对小口相对配置的角接触球轴承，它们只承受径向载荷，因此轴承外圈不需要定位。该主轴选择的轴承类型和配置形式满足主轴高转速和承受较大轴向载荷的要求。主轴受热变形向后伸长，但不影响加工精度。

（2）Tool clamping device　The tool clamping device locates at the lower part of the spindle. Figure 6-26 is a zoomed in view of this part. It consists of a pulling bar, steel balls, a disc spring, a piston and a cylinder. The actions of the tool clamping device during tool changing are shown in following flowchart（see Figure 6-27）

刀具自动夹紧机构位于主轴内部和后端，主要由拉杆、拉杆端部的四个钢球、碟形弹簧、活塞和液压缸等组成。换刀动作如图6-27所示。

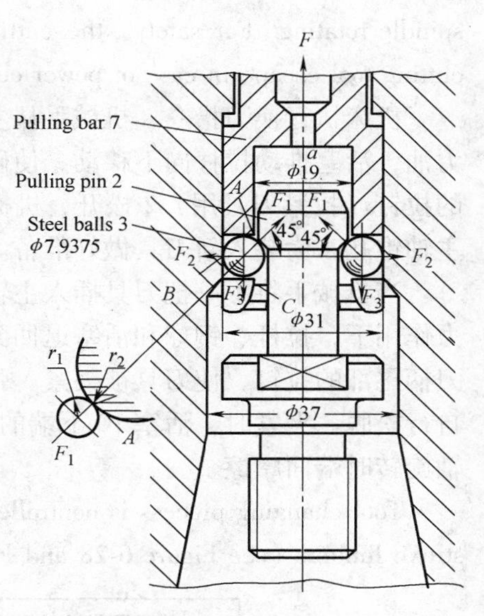

Figure 6-26　Cutting tool clamping mechanism（zoomed in）

The gap between the piston（10）and the pulling bar（7）is designed to avoid end surfaces abrasion during

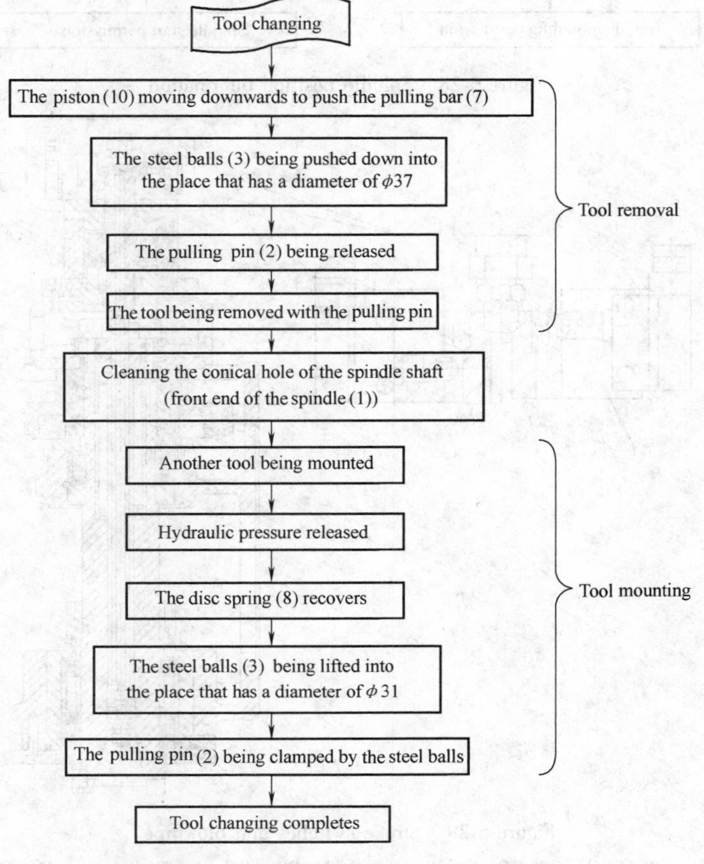

Figure 6-27　The actions of tool clamping device

spindle rotating. For safety, the cutting tool is clamped by the spring. It eliminates the risk of cutting tool escape in case of power cut.

机床执行换刀指令，机械手从主轴拔刀时，主轴需松开刀具。这时，液压缸上腔通入压力油，活塞推动拉杆向下移动，使碟形弹簧压缩，钢球进入主轴锥孔上端的槽内，刀柄尾部的拉钉（拉紧刀具用）2 松开，机械手拔刀。之后，压缩空气进入活塞和拉杆的中孔，吹净主轴锥孔，为装入新刀具做好准备。

当机械手将下一把刀具插入主轴后，液压缸上腔无油压，在碟形弹簧 8 和弹簧 9 的回复力作用下，拉杆、钢球和活塞退回到图 6-27 所示的位置，即碟形弹簧通过拉杆和钢球拉紧刀柄尾部的拉钉，使刀具被夹紧。弹簧夹紧的好处是能保证在工作中突然停电时，刀杆不会自行松脱。夹紧时，活塞 10 下端的活塞杆端与拉杆 7 的上端部之间有一定的间隙，以防主轴旋转时端面摩擦。

Tool changing process is controlled by PLC. The spindle position information is detected by the stroke limiters（see Figure 6-28 and Figure 6-29）.

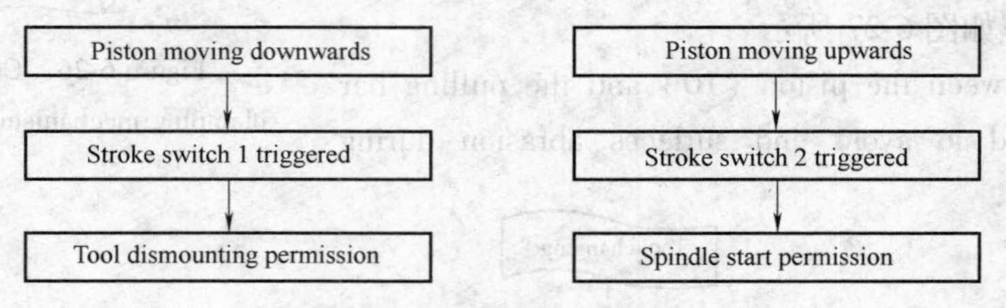

Figure 6-28 Spindle position information

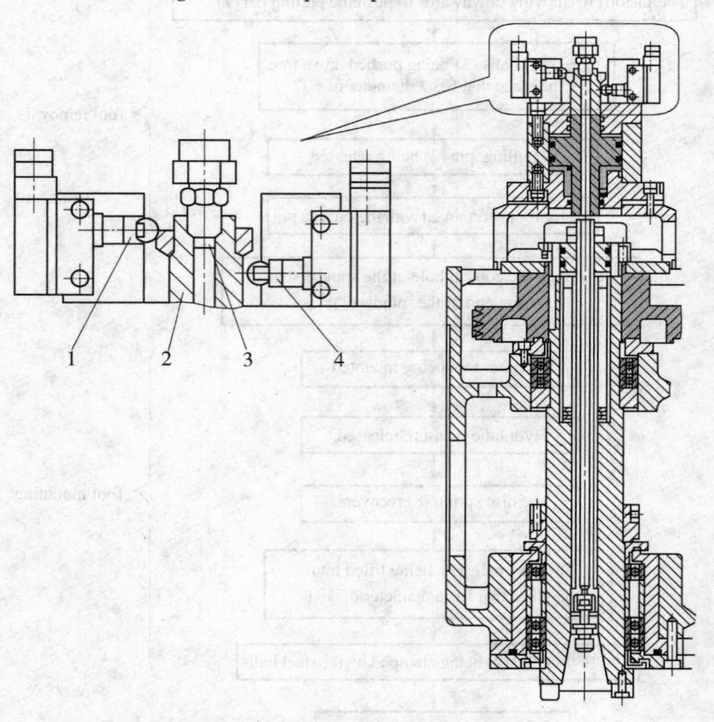

Figure 6-29 Stroke switches and blowpipe

1—Stroke switch 2 2—Pistion bar 3—Blowpipe 4—Stroke switch 1

(3) Cleaning blowpipe Chips, dust or other contaminants in the conical hole of the spindle will abrade the tool handle thus affecting tool positioning accuracy. Therefore, when cutting tool is dismounted, it is necessary to blow compressed air through the center holes of the piston and the pulling bar to clean the conical hole.

换刀过程中，如果主轴锥孔内落入了切屑、灰尘或其他污物，在拉紧刀杆时，锥孔表面和刀杆锥柄会受到损伤，破坏定位。为了保持主轴锥孔的清洁，刀具拔出后，压缩空气经活塞由孔内的空气嘴喷出，将锥孔清理干净。

(4) Spindle orientation device The cutting torque is delivered by keys on the end plane of the spindle and the keyway on the end surface of the cutting tool. During tool mounting, the key and the keyway must be precisely oriented. The JCS-018A is equipped with an electromagnetic orientation device. Refer Chapter 3. 1. 4 for the principle of the orientation.

机床的切削转矩由主轴上的端面键来传递，每次机械手自动装取刀具时，必须保证刀柄上的键槽对准主轴的端面键。本机床采用的是电气式主轴准停装置，即用磁力传感器检测定向，其工作原理参看本书3. 1. 4。

2. Feeding servo system

(1) Feeding servo systems for *X*-, *Y*-, and *Z*-axis Feeding servo system for *X*-, *Y*-, and *Z*-axis have a similar construction, which is shown in Figure 6-30. The system is driven by a pulse width controlled the DC motor. The connection between the motor shaft and the left half of the coupling (2) is realized by a group of conical rings instead of a key to eliminate transmission clearance. The right half of the coupling (2) connects to the ball screw (3) by a key. The nut unit consists of two halves (4) and (7) for clearance and preloading force adjusting. When the ball screw (3) rotates, the nut unit performs axial motion with the worktable.

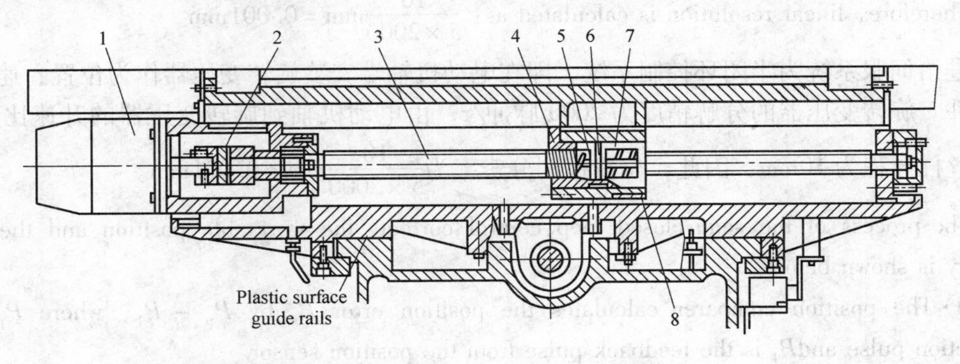

Figure 6-30 A feeding servo assembly

1—DC servo motor 2—Oldham's coupling 3—Ball screw

4—Left half of the nut 5—Key 6—Open washer 7—Right half of the nut 8—Nut housing

The main difference in *Z*-axis feeding servo system is the brake. When the servo motor stops, an electromagnetic switch releases the spring. Without the compressive force, the spring then immediately recovers to engage frictional discs to stop the worktable from moving.

机床有三套（*X*、*Y*、*Z* 轴）相同的伺服进给系统，如图 6-30 所示。该系统由脉宽调速直流伺服电动机 1 驱动，采用无键连接方式，用锁紧环将运动传至滑块联轴器 2 的左连接件。联轴器的右连接件与滚珠丝杠 3 用键相连，由滚珠丝杠 3、左螺母 4 和右螺母 7 驱动工作台移动。滚珠螺母由左螺母 4 和右螺母 7 组成，并固定在工作台上。

由于滚珠丝杠没有自锁能力，在 Z 向进给伺服电动机上加有制动装置。当电动机停转时，电磁线圈的电流被切断，松开弹簧，由弹簧压紧摩擦片使其制动。

（2）Feeding servo control system （see Figure 6-31） Motion speed and position are monitored by the rotary transformer that is fixed on the end of the motor shaft（so it is a semi-closed loop control）. The rotary transformer has an angular resolution of 2, 000 pulses/rev. Speedup rate is 5:1 from the motor shaft to the rotary transformer. The ball screw has a pitch of 10mm.

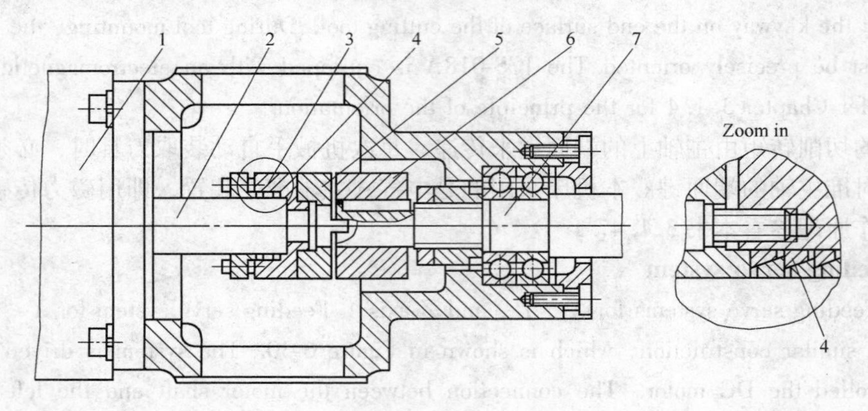

Figure 6-31 Connection between servo motor and ball screw

1—DC servo motor 2—Motor shaft 3—Sleeve

4—Conical rings 5—Oldham's coupling 6—Sleeve 7—Ball screw

Therefore, linear resolution is calculated as: $\frac{10}{5\times2000}\text{mm}=0.001\text{mm}$

进给伺服系统为半闭环控制系统，即在电动机轴端安装旋转变压器作为位置、速度的反馈元件。旋转变压器的分解精度为2000 脉冲/r，由电动机轴到旋转变压器的升速比为 5:1，滚珠丝杠导程为10mm，因此，位置检测分辨率为$\frac{10}{5\times2000}\text{mm}=0.001\text{mm}$。

The process of the semi-closed loop control corrects the worktable position and the motion velocity is shown below:

① The position comparer calculates the position error P_e by $P_p - P_1$, where P_p is the instruction pulse andP_1 is the feedback pulse from the position sensor.

② The digit-analog signal converter converts the error P_e into an analog voltage U_e.

③ Signal amplifier amplifies the voltage U_e to U_c.

④ The velocity comparer calculates velocity error U_a by $U_c - U_g$, where U_g is an analog voltage produced by velocity sensor.

⑤ Figure 6-32 expresses the signal processing steps of ① ~ ⑤ by a flowchart. The velocity error U_a then is amplified to U_m to drive the servo motor.

从计算机来的位置指令脉冲 P_p，在位置偏差检测器内与位置检测器送来的反馈脉冲 P_1 比较，其差值为 P_e，经数-模转换器转换为差值的模拟电压 U_e。然后，位置控制放大器把 U_e

放大为 U_c，送至速度误差检测器与速度检测器来的速度（转速）模拟电压 U_g 比较，其差值 U_a 经速度放大器放大为 U_m 后，对伺服电动机转速进行控制。

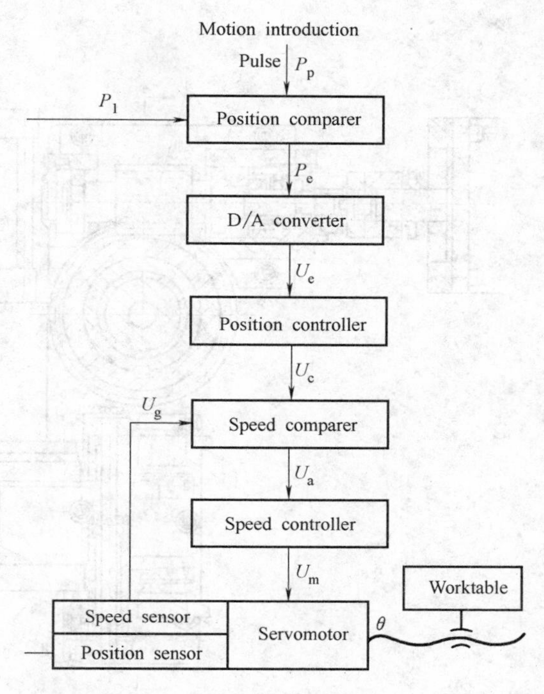

Figure 6-32 Diagram of feeding servo control system

3. Automatic tool changer（ATC）

The ATC of the JCS-18A consists of a magazine and a manipulator. The construction of the magazine drive mechanism is shown in Figure 6-33. The actions of the magazine components during tool selection are shown in Figure 6-34.

自动换刀装置由刀库和机械手两部分组成，刀库的驱动机构如图 6-33 所示。数控系统发出换刀指令后，直流伺服电动机 1 接通，其运动经过滑块联轴器 2、蜗杆 4、蜗轮 3 传到刀盘 14，刀盘带动其上的 16 个刀套 13 转动，来完成选刀工作。每个刀套尾部有一个滚子 11，当待换刀具转到换刀位置时，滚子 11 进入拨叉 7 的槽内。同时，气缸 5 的下腔通压缩空气，活塞杆 6 带动拨叉 7 上升，放开位置开关 9，用以断开相关的电路，防止刀库、主轴等有误动作。拨叉 7 在上升的过程中，带动刀套绕着销轴 12 逆时针向下旋转 90°，从而使刀具轴线与主轴轴线平行。

As shown in Figure 6-35, the axes of cutting tools in the magazine are vertical to the spindle axis. The manipulator cannot catch the cutting tool in such a position. Therefore, when a cutting tool is selected or being replaced into the magazine, the tool holder for the cutting tool must be parallel to the spindle axis.

The manipulator actions consist of tool catching, tools removal, tools exchange, tools mounting, and system reset. These actions are realized by rotating and axial moving of the manipulator. The actions must be performed in sequence, which is controlled by mechanical mechanisms and PLC. The sequence of motions in the process of tool changing is like a "chain reaction."

Figure 6-36 shows the constructions of drive mechanism of the manipulator. Those dashed lines indicate the mechanical or electrical linkage between components. The sequence of components' motions is illustrated in Figure 6-37.

图 6-36 所示为换刀机械手驱动机构的结构示意图。换刀机械手的回转、轴向移动等动作通过各功能部件顺序执行实现，其工作过程可由图 6-37 来表示。

The claws of the manipulator direct contact the cutting tool. The tool is expected to be firmly held during motion, and can be caught or released conveniently.

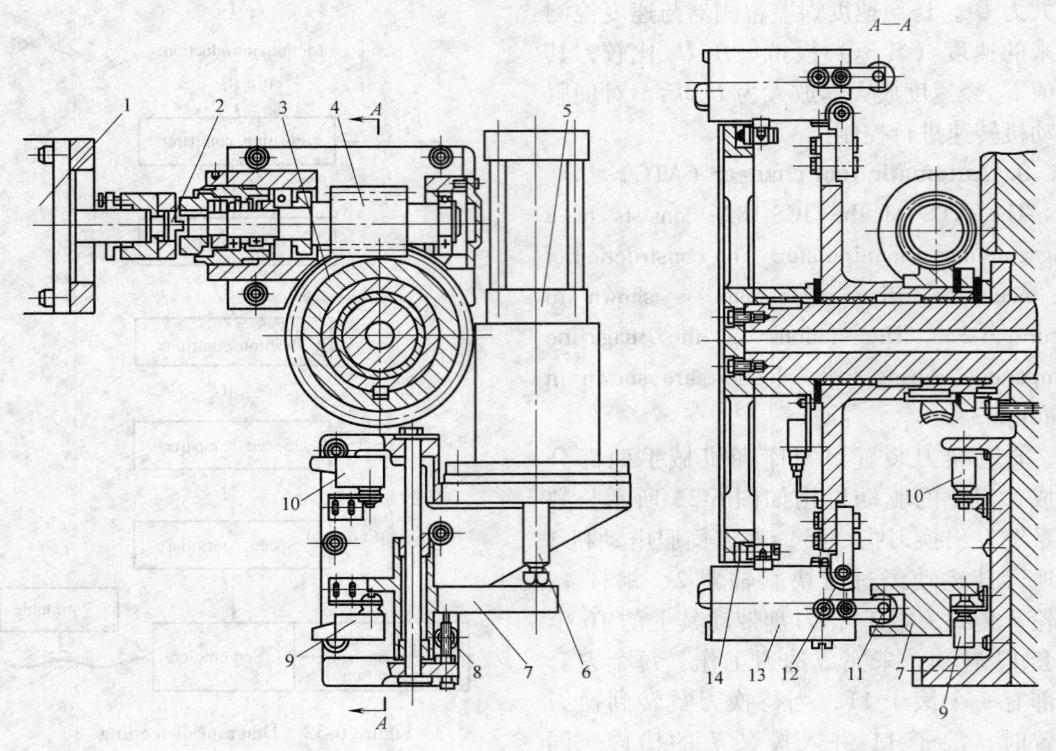

Figure 6-33 Construction of tool magazine driving mechanism

1—DC servo motor 2—Oldham's coupling 3—Worm gear 4—Worm shaft
5—Cylinder 6—Piston bar 7—Fork 8—Adjusting screw 9—Stroke switch 1
10—Stroke switch 2 11—Roller 12—Pin 13—Tool holder 14—Tool plate

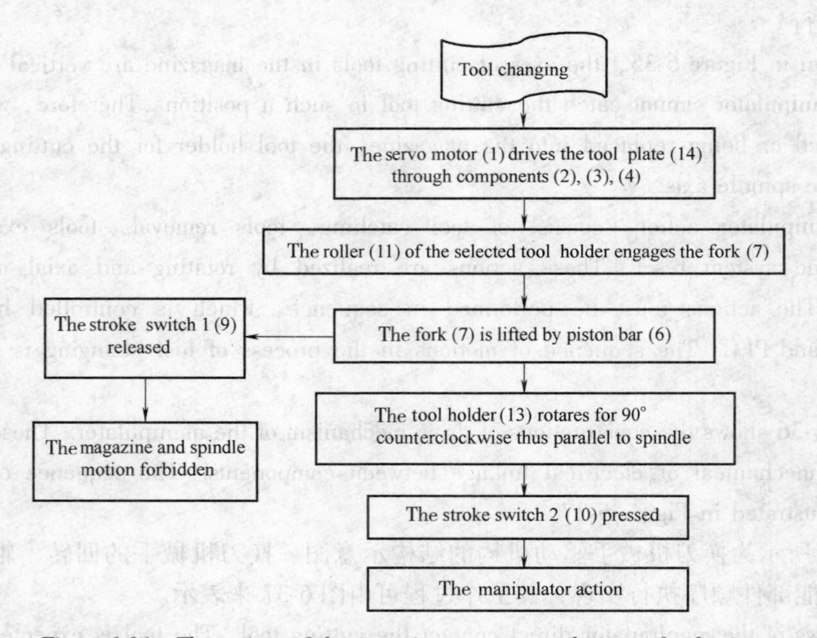

Figure 6-34 The actions of the magazine components during tool selection

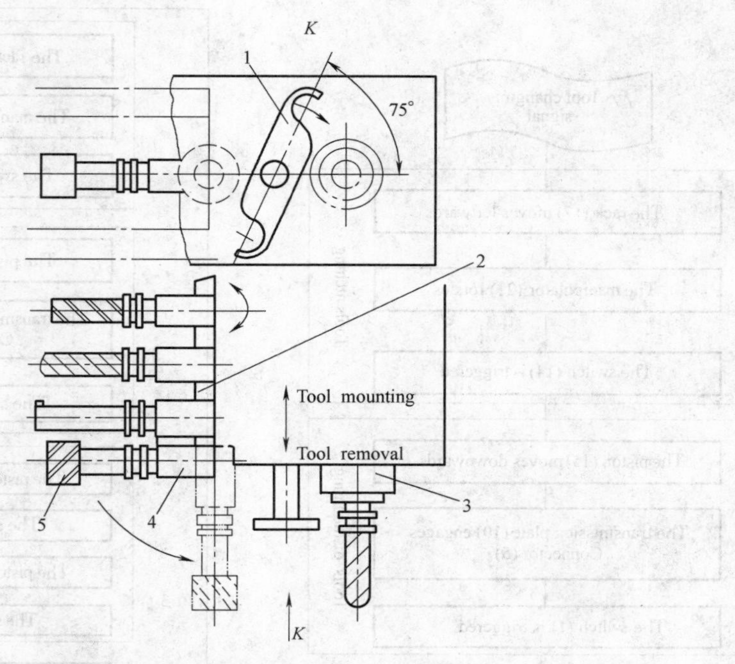

Figure 6-35 Tool holder position during tool exchange

1—Manipulator 2—Magazine 3—Spindle 4—Tool holder 5—Cutting tool

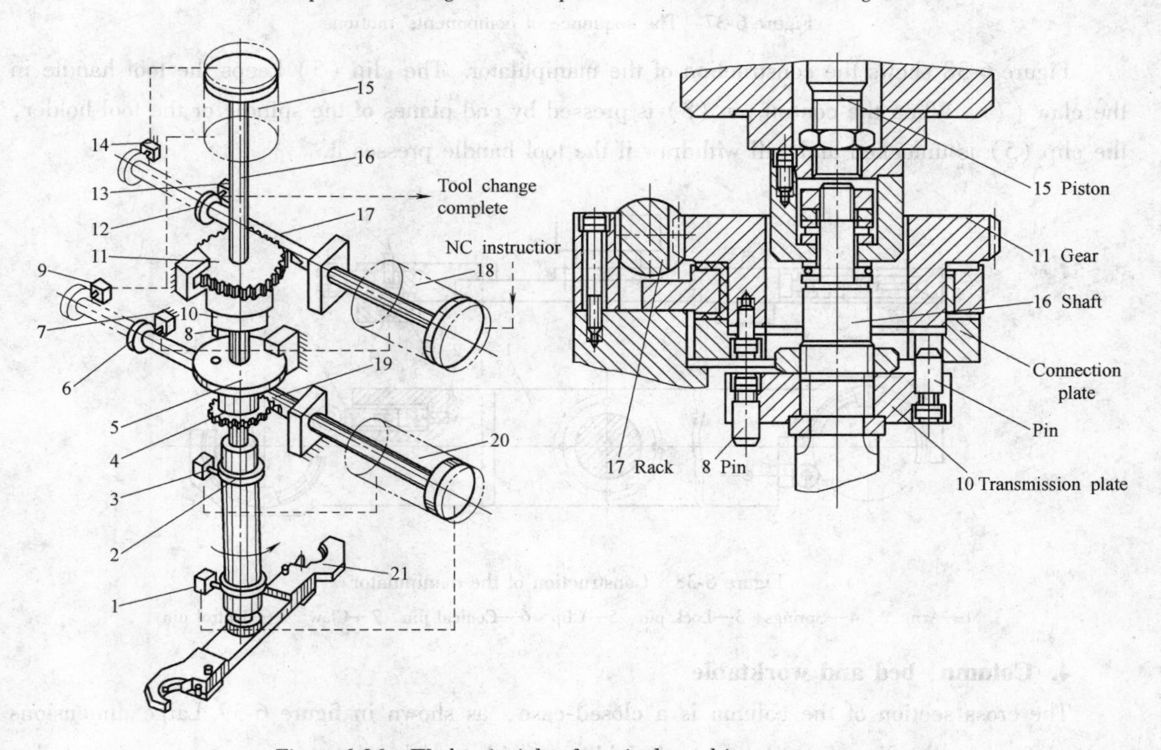

Figure 6-36 Work principle of manipulator drive components

1、3、7、9、13、14—Limit switches 2、6、12—Rings 4、11—Gears 5—Connector 8—Pin

10—Transmission plate 15、18、20—Piston 16—Shaft 17、19—Racks 21—Manipulator

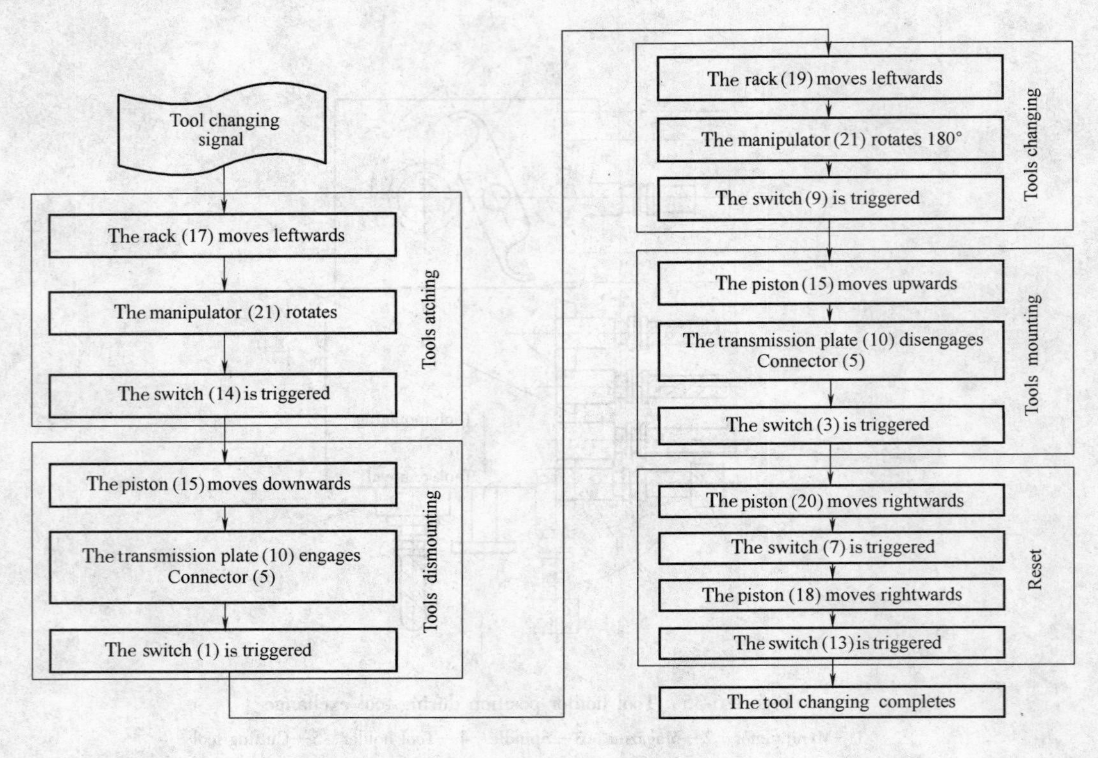

Figure 6-37 The sequence of components' motions

Figure 6-38 shows the construction of the manipulator. The clip (5) keeps the tool handle in the claw (7). When the control pin (8) is pressed by end planes of the spindle or the tool holder, the clip (5) is unlocked and will withdraw if the tool handle presses it.

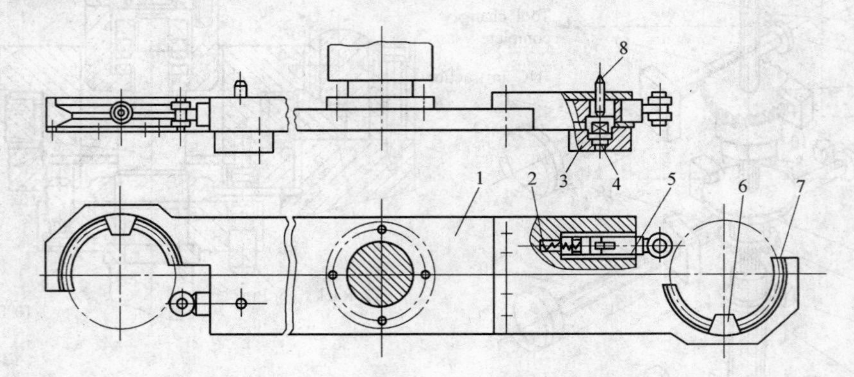

Figure 6-38 Construction of the manipulator

1—Arn 2、4—Springs 3—Lock pin 5—Clip 6—Conical pin 7—Claw 8—Control pin

4. Column, bed and worktable

The cross section of the column is a closed-case, as shown in figure 6-39. Large dimensions and reinforcing ribs give the column good rigidity to withstand bending moment and torsion during cutting.

本机床的立柱为封闭的箱型结构，承受弯矩和扭矩，较大的截面尺寸及内壁的加强肋使

立柱具有较大的刚度。

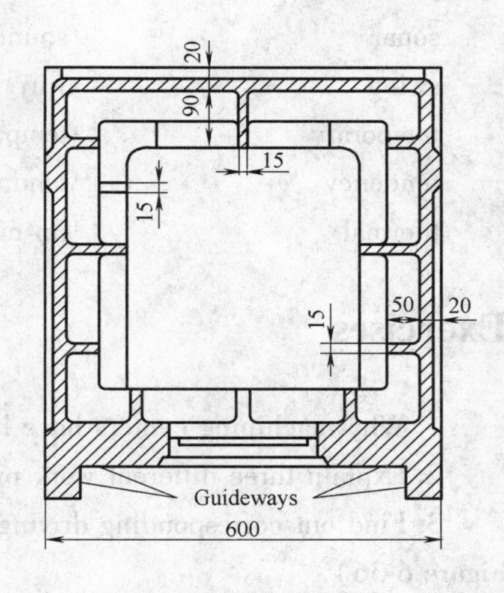

Figure 6-39 Cross section of the column

The JCS-018A applies plastic covered guideways to reduce friction and eliminate creeping at low speed. Since the plastic guideways surface has nice lubricating characteristics, intermittent lubricant supply meets the lubrication requirements. Usually, 1. 5-2. 5mL of lubricant is pumped every 7. 5 minutes.

A check valve and throttle orifices are used to prevent lubricant flowing back from ways to the pipe in case of pump stop.

该机床的导轨面上都贴有氟化乙烯导轨板，摩擦系数小，低速运动时，无“爬行”现象发生。由于氟化乙烯导轨板的润滑性良好，而且对润滑油的供油量要求不高，因此，机床只用了间隙式润滑泵供油。每 7. 5min 泵油一次，每次泵油量为 1. 5 ~ 2. 5mL。润滑点的管接头内有单向阀和节流小孔，防止当油泵停止泵油时导轨间的润滑油被挤回油管。

Glossary

acoustic	[ə'kuːstik]	听觉的；声学的
ATC		自动换刀装置(automatic tool changer)
blowpipe	['bləuˌpaip]	吹风管
claw	[klɔː]	卡爪，钳形器具
complex	['kɔmpleks]	复杂的
concentrated	['kɔnsentreitid]	集中的
controllability	[ˌkən'trəuləbiliti]	可控制性
elevate	['eliveit]	举起；使上升
encoding	[in'kəudiŋ]	编码
gantry	['gæntri]	门架结构；龙门起重架
inspection	[in'spekʃən]	检查；检验
intelligent	[in'telidʒənt]	智能的
magazine	[ˌmægə'ziːn]	（机床的）刀库
manipulator	[mə'nipjuleitə]	机械手
modification	[ˌmɔdifi'keiʃən]	修改；改装
physical	['fizikəl]	物质的
pigeonhole	['pidʒinˌhəul]	鸽舍；分类架
sensor	['sensə]	传感器
sequential	[si'kwenʃəl]	顺次的
significant	[sig'nifikənt]	显著的；重要的

sonar	[ˈsəunɑː]	声纳；音波探测器
swing	[swiŋ]	（在轴上）摆动
temporary	[ˈtempəreri]	临时的，暂时的
tendency	[ˈtendənsi]	倾向，趋势
thermal	[ˈθəːməl]	热量的

Exercises

1. Why machining centers have high machining efficiency?

2. Explain three different work principles of cutting tool identification.

3. Find out corresponding driving components of manipulator motions during tool changing (see Figure 6-36).

4. A worktable transmission chain is shown below:

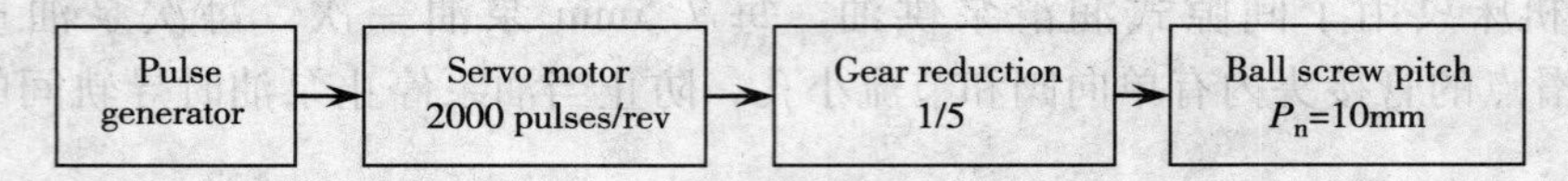

(a) Determine the pulse equivalent of the system.

(b) Determine the motion speed of the worktable if the servo motor receives 800, 000 pulses/min.

Chapter 7　EDM Machine Tools

7.1　Sinker EDM machine

7.1.1　Work principle of sinker EDM

The basic principle of sinker EDM is "electrical erosion", which is known as removing metal from the workpiece surface by electrical spark thus obtaining certain shape, dimensions, surface qualities of a workpiece. Figure 7-1 shows sinker EDM processing (left) and tool electrodes (right).

Figure 7-1　Sinker EDM processing (left) and tool electrodes (right)

电火花加工建立在"电蚀"现象基础上，通过在一定介质中的工具电极和工件之间脉冲性火花放电的电腐蚀作用来蚀除多余的金属，从而获得所需零件的尺寸、形状及表面质量。图7-1所示为电火花加工过程（左图）和各种工具电极（右图）。

Figure 7-2 shows the work principle of the sinker EDM machine tool. As the power source, the pulse generator (2) produces electrical pulses on the workpiece and the tool electrode (4). The feeding mechanism (3) keeps a certain gap between the tool and the workpiece to maintain continuous machining.

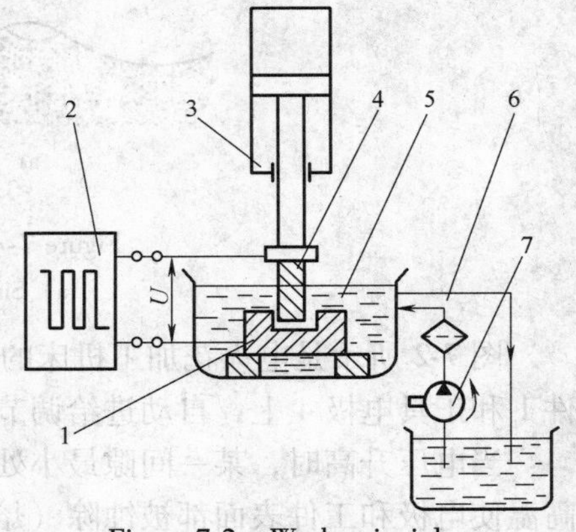

Figure 7-2　Work principle of the sinker EDM machine tool

1—Workpiece　2—Pulse generator　3—Feeding mechanism　4—Tool electrode　5—Coolant　6—Filter　7—Pump

When the voltage between the workpiece and the tool electrode rises, the work fluid (insulative coolant) at the narrowest gap will be electrically penetrated. An arc then produces instantaneous high temperature (8,000-12,000℃) to melt or vaporize the metal both on the workpiece and the tool electrode nearby. The process is shown in Figure 7-3.

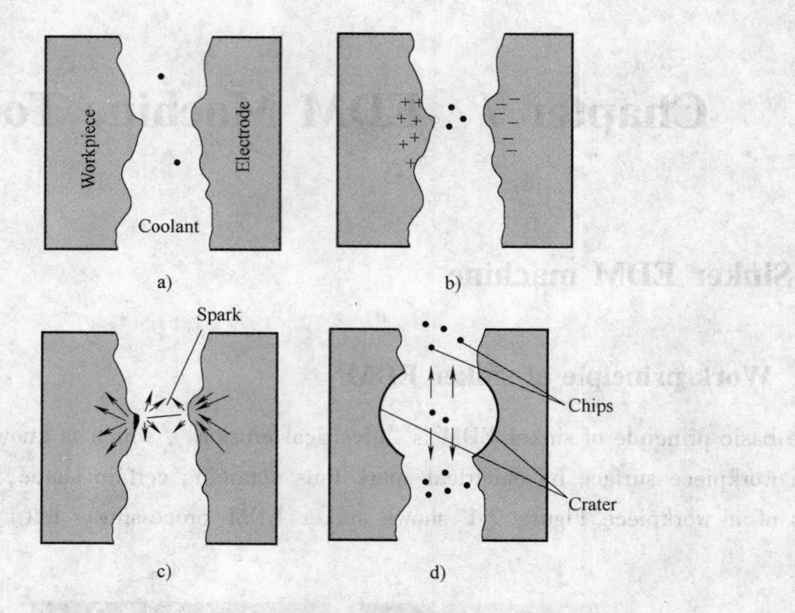

Figure 7-3 Stages of EDM process

a) Insulation by coolant b) Electrlc fleld concentration

c) Electrical spark discharge d) Recovery of insulation

Machining speed of the EDM depends on energy of each spark. Material removal rate ranges from one to thousands mm^3 per minute. Energetic sparks result in fast machining but rough surface (see Figure 7-4a). In contrast, high frequency but weak energy sparks produce fine surface in low efficiency (see Figure 7-4b).

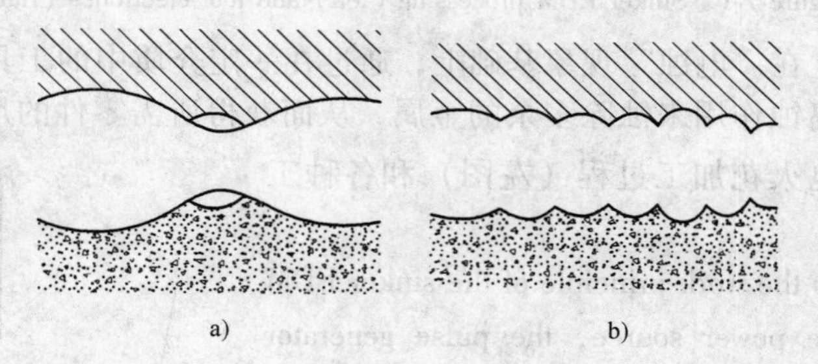

Figure 7-4 Surfaces after electrical erosion

a) Single pulse b) Continuous pulses

图 7-2 所示是电火花加工机床的原理图。由脉冲电源 2 输出的电压加在液体介质中的工件 1 和工具电极 4 上，自动进给调节装置 3 使电极和工件间保持一定的放电间隙。

当电压升高时，某一间隙最小处或绝缘强度最低处的介质被击穿，产生火花放电，瞬时高温使电极和工件表面都被蚀除（熔化或汽化）一小块材料，各自形成一个小凹坑。

7.1.2 Features of EDM

1. Advantages of EDM

1) Tool material hardness is no longer a barrier of machining since ultra-high temperature of

the electrical arc is able to melt or vaporize any material. Therefore, EDM is suitable for machining any conductive material regardless of hardness, brittleness, toughness（韧度）and melting point.

2）There is no cutting force during machining since no mechanical contact between the electrode and the workpiece, therefore, it is suitable for machining those workpieces that have special structure, e. g. thin-wall, narrow slot, non-round holes, etc.

3）Rough and finish machining can be continually performed by different pulse width on the same EDM machine tool.

4）Since the machining directly utilizes electric energy, automatic control and automatic machining can be easily realized.

5）Reduce risk of hazardous dust particles from material since particles are flushed away to the filter（see Figure 7-5）.

电火花加工的优点：

1）脉冲放电产生的高温足以熔化和汽化任何材料，因此电火花加工适合加工用传统机械加工方法难加工的特殊材料，如各种淬火钢、硬质合金、耐热合金等任何硬、脆、韧、高熔点的导电材料。

2）电火花加工时工具电极与工件不接触，两者之间不存在因切削力而产生的一系列设备和工艺问题，因此不受工件几何形状的影响，适合加工形状复杂和有特殊工艺要求的工件，如薄壁、窄槽、各种型孔、立体曲面等。

3）脉冲宽度可以调节，在同一台机床上能连续进行粗、精加工。

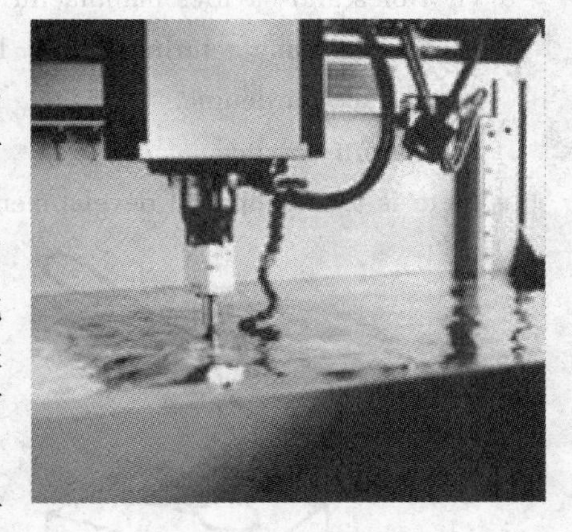
Figure 7-5　Machining under fluid

4）直接利用电能加工，便于实现自动控制和加工自动化。

5）加工在工作液中进行，因此金属碎屑直接被冲入过滤器，不会污染环境，也不会对操作人员的健康产生危害。

2. Shortages of EDM

1）The tool electrode must be made first. It is inconvenient since the tool electrode making is expensive and time consuming.

2）Machining area may be unfavorably hardened. Though machining area is small, hardened layer due to fast cooling after heating by ultra high temperature may crack.

3）Discharge gap produces error. The gap must exist for electrical arc generating. However, the gap affects machining accuracy.

4）Machining accuracy drops with electrode wearing. Eroded electrode needs repair or replacement.

Though EDM has many disadvantages, it is more suitable for metal cutting compared to conventional machine tools

电火花加工的缺点：

1）电火花加工必须要有工具电极，与其他加工方法相比，增加了制作电极的费用和时间。

2）工件上进行电加工的表面由于受高温加热后急速冷却，会生成电加工表面变质层，容易造成加工部位的碎裂与崩刃。

3）电极和工件之间需有一定的加工间隙，使得加工误差增大。

4）电极在加工过程中受到电蚀损耗会影响加工精度。损耗的电极需要进行修整或更换。

尽管如此，由于电火花加工具有许多其他加工方法无可替代的优点，因此仍为一种应用广泛的金属加工方法。

7.1.3 Application of sinker EDM in mold manufacturing

In mould industry, EDM has wide application for its convenience. Today, about 70% of machining processes are performed by EDM in mould manufacturing. For mould industry, EDM is applied to:

① Holes and cavities manufacturing.

② Orifice manufacturing ($d > 0.1$mm).

③ Surface hardening.

④ Pattern carving.

Here is an example of development of a plastic injection mould of a spoon by EDM.

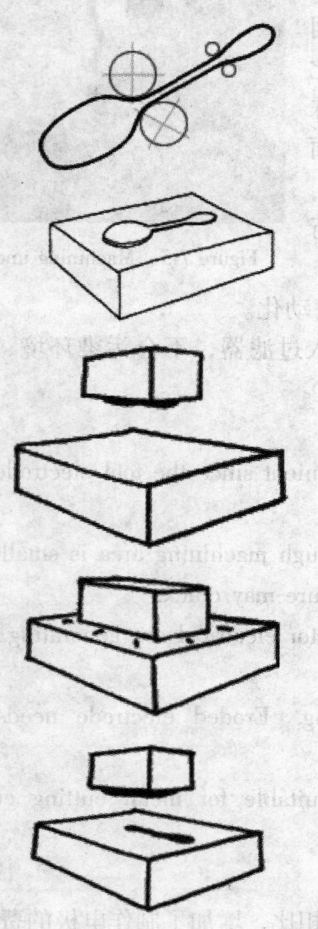

Step 1

The design has to be worked out, and the shape of the spoon is reproduced on a copper or graphite electrode.

Step 2

The electrode is then mounted on an EDM machine, together with the workpiece in steel that is going to become the plastic injection mold.

Step 3

The electrode will penetrate into the workpiece by electrical discharge erosion.

Step 4

Create a negative impression of the spoon.

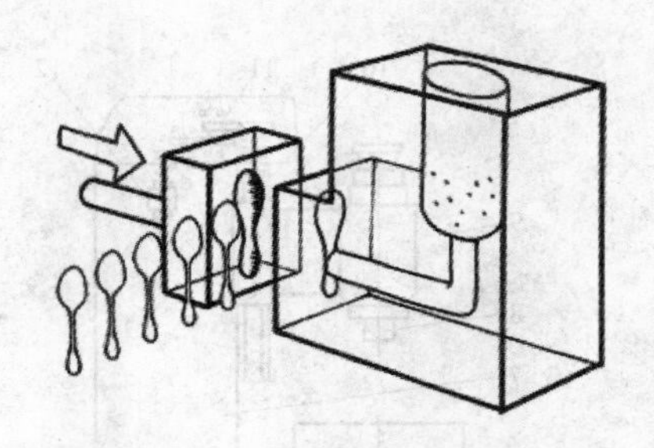

Step 5

Finally the workpiece with the impression is mounted on a plastic injection machine. After everything is set up, mass production could be started.

电火花加工在模具制造行业中获得了广泛的应用，模具加工约70%的工作量可由电火花加工来完成。电火花加工在模具制造中的应用主要有：

① 加工各种模具零件的复杂型孔和型腔。

② 加工各种微小的圆孔、异形孔。

③ 强化金属表面，提高耐用度。

④ 图形蚀刻。

7.1.4 Basic construction of a sinker EDM machine

A sinker EDM machine consists of four systems, namely, machine tool reality, pulse generator (power unit), feeding/adjusting system, and work fluid circulation and filtration system. Figure 7-6 shows the basic construction of a sinker EDM machine.

电火花加工机床的基本结构如图7-6所示，它由机床本体、脉冲电源、自动调节系统和工作液循环过滤系统四部分组成。

1. Machine tool reality

The machine tool reality consists of a bed, a column, a worktable and a spindle head. It is the framework ensuring accurate relative position between the tool electrode and the workpiece.

Figure 7-7 shows a typical machine tool configuration and the mechanical transmission system of a sinker EDM machine.

机床本体的作用是保证工具电极与工件之间的准确的相对位置，它主要包括床身、立柱、工作台和主轴头。图7-7所示是典型的电火花加工机床的机床本体和机械传动系统。

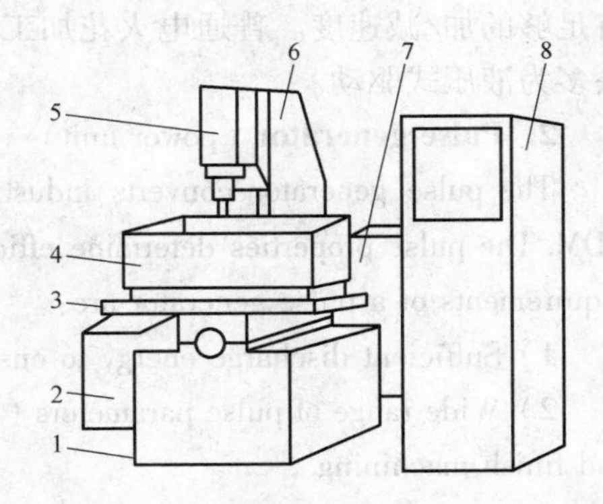

Figure 7-6 Basic construction of sinker EDM machine

1—Bed 2—Hydraulic reservoir 3—Worktable 4—Work fluid sink 5—Spindle head 6—Column 7—Work fluid tank 8—Power unit

(1) Column and bed As the basic components, the column and the bed are expected to have enough rigidity. Accurate perpendicularity between the bed and the column, and between the worktable and the column are required.

床身和立柱是机床的主要基础件，要有足够的刚度。床身、工作台与立柱导轨之间有一定的垂直度要求。它们的刚度、精度和耐磨性对电火花加工质量有直接影响。

(2) Worktable The worktable holds a work fluid sink, and clamps the workpiece. It is driven

by handwheels through ball screw-nut transmission to perform motion in X-Y plane.

工作台支承和安装工作液槽及工件。通过转动纵向、横向手轮，带动丝杠来移动工作台，从而改变工具电极和工具的位置。

（3） Spindle head　The spindle head is the actuator of the automatic adjusting system that controls the gap between the electrode and the workpiece. As a key component of a sinker EDM machine, the spindle head is required to have simple structure for reliability, short transmission chain for accurate motion, and sufficient acceleration/deceleration of motion for rapid response of feeding and withdrawing. Spindle heads of most sinker EDM machines are hydraulically driven.

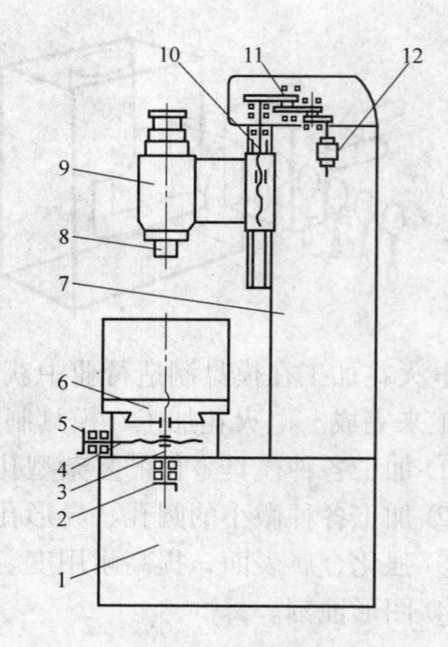

Figure 7-7　Machine tool reality and mechanical transmission system of a sinker EDM machine

1—Bed　2、5—Handwheels　3、4、10—Ball screws　6—Worktable　7—Column　8—Spindle head　9—Spindle case　11—Gear set　12—Motor

主轴头是电火花加工机床的一个关键部件，它是自动调节系统的执行机构，控制工具电极与工件的间隙。主轴头要求结构简单，传动链短，有足够的加/减速度。普通电火花加工机床的主轴头多为液压式驱动。

2. Pulse generator（power unit）

The pulse generator converts industrial AC to unilateral pulse current to supply energy for EDM. The pulse properties determine efficiency, stability, accuracy and roughness of EDM. Basic requirements of a pulse generator are:

1）Sufficient discharge energy to ensure metal melting or vaporization instantaneously.

2）Wide range of pulse parameters（e. g. pulse width, pulse intervals, peak current）for rough and finish machining.

3）Durable tool electrode. Wearing（erosion）should be less than 0.5% of that of the workpiece.

4）Steep edges of the pulse. The rectangular wave pulse is usually applied for steep edge . It can minimize the impacts of the gap and contaminants between the electrode and the workpiece.

5）Reliable characteristics, simple construction, convenient operation and maintenance.

电火花加工机床脉冲电源的作用是将工频交流电转变成一定频率的单向脉冲电流，以提供电火花加工所需要的能量。脉冲电源的性能直接影响到电火花加工的生产效率、加工稳定性、加工精度和表面粗糙度。因此，对脉冲电源有以下基本要求：

1）脉冲放电能量足够，保证金属能被瞬时熔化或汽化。

2）脉冲电源的脉冲宽度、脉冲间隙和峰值电流应该有较宽的调节范围，以满足粗、精加工的需要。

3）工具电极损耗要小，粗加工时相对损耗要小于0.5%。

4）一般常采用矩形波脉冲电源，因为矩形波脉冲前后沿较陡，能减少电极间隙的变化及油污程度等对脉冲放电宽度和能量等参数的影响，稳定工艺过程。

5）性能稳定可靠，结构简单，操作和维修方便。

3. Automatic feeding/adjusting system

A certain gap between the workpiece and the tool electrode ensures EDM to take place. The dimension of the discharge gap depends on machining requirements. With the machining process, continuous electrical erosion expands the gap until the electrical discharge ceases. To remain the discharge, the tool electrode must perform feeding motion. In addition, when chips accumulate and result in short circuit between the workpiece and the tool electrode, the automatic adjusting system keeps the gap at a constant value for optimum electrical discharge. The automatic feeding/adjusting system is usually driven by electric-hydraulic system or a servo motor. Figure 7-8 shows the discharge clearance.

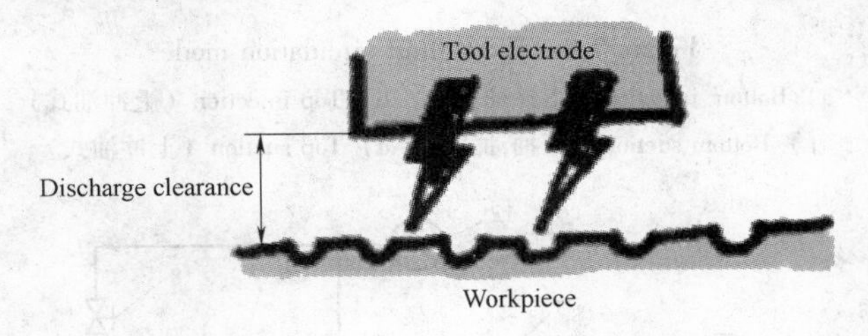

Figure 7-8 Discharge clearance

电火花加工时，工件与电极之间必须要保持一定的距离才能维持放电，这段距离称为放电间隙（图7-8）。放电间隙取决于粗、精加工所选用的电参数。工件和电极随着加工过程都有一定的损耗，这就使得放电间隙逐渐增大，当间隙大到不足以维持放电时，加工便停止。为了使加工能继续进行，电极必须不断地、及时地进给，以维持所需的放电间隙。当外部的干扰使放电间隙发生变化（如排屑不良而造成短路）时，调节系统也会随之自动作相应的变化，以保持最佳放电间隙。目前自动调节系统多为电液驱动和伺服电动机驱动。

4. Circulation and filtration system of work fluid

The circulation and filtration system of a sinker EDM machine tool consists of a pump, fluid containers, a filter and pipes. It propels work fluid circulation.

The work fluid pressure can be adjusted between 0-200kPa, depending on machining depth and contour. Fluid flow flushes chips between the electrode and the workpiece away to obtain a stable machining process.

Figure 7-9 shows four modes of work fluid circulation. Though fluid suction mode provides a cleaner discharge environment, injection mode has wider application.

电火花加工用的工作液循环过滤系统包括工作液泵、容器、过滤器及管道等，它能使工作液进行强制循环。加工时，工作液的压力可根据工件的几何形状及加工的深度进行调整，调整范围一般为0～200kPa。图7-9所示为四种工作液循环方式。冲油方式的循环效果比抽油式的更好，特别是在型腔加工中，采用这种方式可以改善加工的稳定性。

Figure 7-10 shows the circuit of fluid circulation. It can operate in both injection and suction modes by switch valve selection. The work principle of the system is illustrated by Figure 7-11.

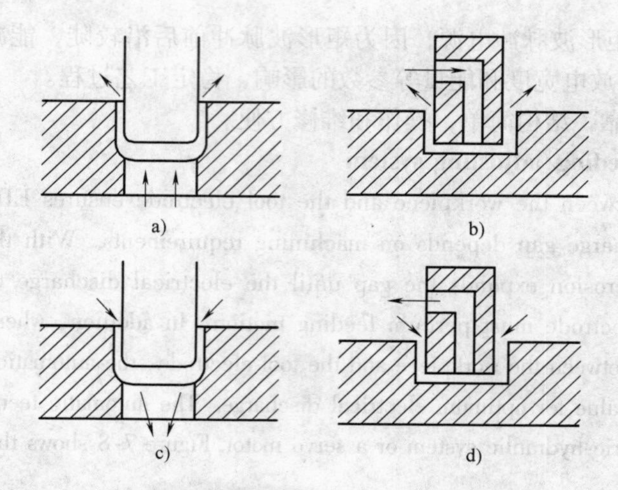

Figure 7-9　Work fluid circulation mode

a）Bottom injection（下冲油式）　b）Top injection（上冲油式）

c）Bottom suction（下抽油式）　d）Top suction（上冲油式）

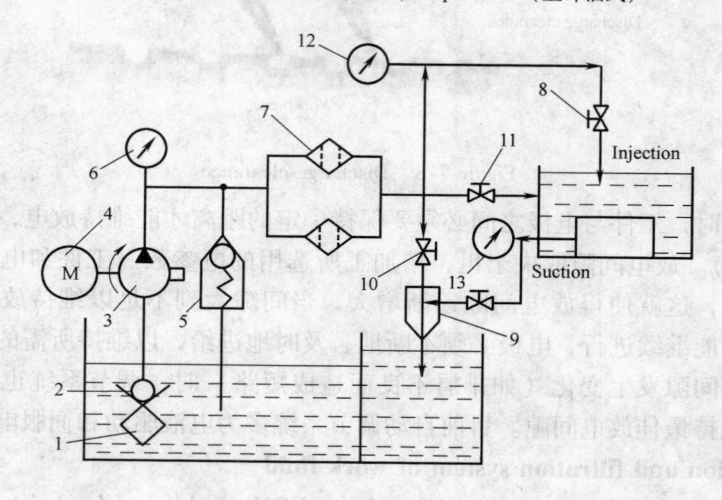

Figure 7-10　Work fluid circulation mode

1—Coarse filter　2—Single-direction valve　3—Pump　4—Motor

5—Safety valve　6—Pressure meter　7—Fine filter　8—Pressure adjusting valve

9—Suction pipe　10—Injection/suction switch valve　11—Flow Control valve

12—Injection pressure meter　13—Injection pressure meter

图 7-10 所示是工作液循环系统油路，既能实现冲油，又能实现抽油。其工作过程是：储油箱的工作液首先经粗过滤器 1、单向阀 2 吸入泵 3，这时高压油经过精过滤器 7 输向机床工作液槽。安全阀 5 控制系统的压力不超过 400kPa。流量控制阀 11 为快速进油用，待油注满油箱时，可及时调节冲抽/油开关阀 10，由压力调节阀 8 来控制工作液的循环方式及压力，当冲抽/油开关阀 10 在冲油位置时，补油和抽油都不通，这时油杯中油的压力由压力调节阀 8 控制。当冲抽/油开关阀 10 在抽油位置时，补油和抽油两路都通，这时压力工作液穿过射流吸管 9，利用流体速度产生负压，达到抽油的目的。

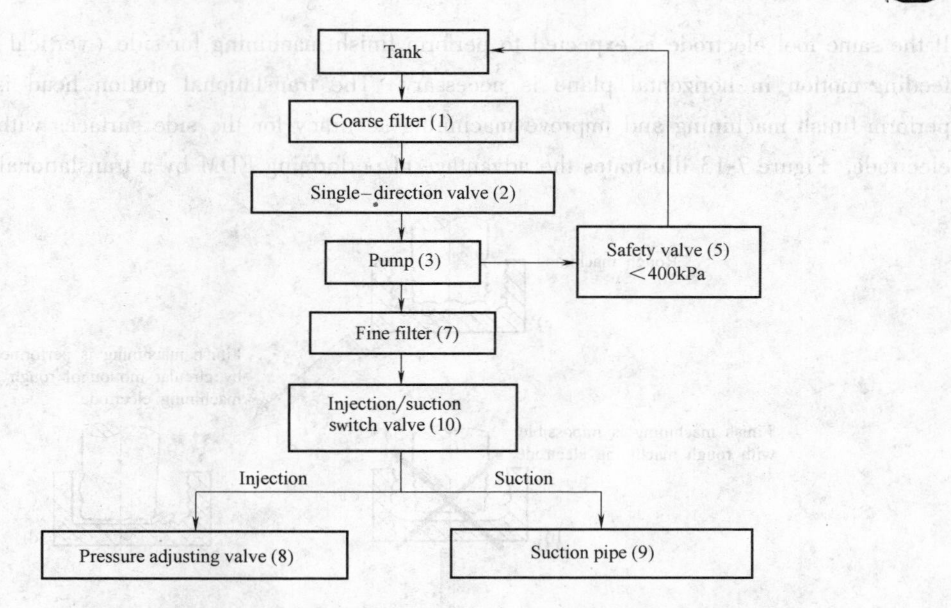

Figure 7-11 Work principle of the circulation and filtration system

7.1.5 Accessories of EDM machines

1. Position-adjustable electrode collet

Before machining, the axis of tool electrode on the spindle head must be adjusted to be perpendicular to the datum plane of the workpiece. It is realized by position adjusting mechanism of the collet (see Figure 7-12). The adjusting is a manual operation. The collet with a tool electrode should be insulated from other mechanical components of the machine tool.

主轴上的工具电极在加工前需要调节到与工件基准面垂直，可通过夹头的调节机构（图7-12）来实现。垂直度与水平转角调节正确后，分别用螺钉卡紧。此外，机床主轴、床身在电路上连成一体接地，而工具电极的夹持调节部分应单独绝缘。

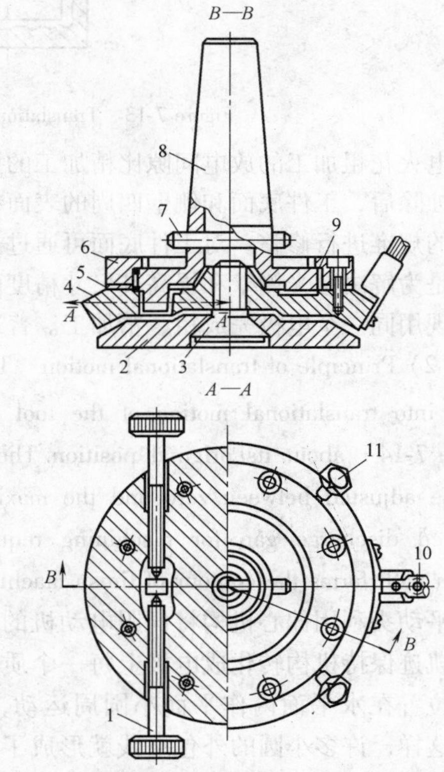

Figure 7-12 Collet and position adjusting mechanism

1—Adjusting screw 2—Swing flange 3—Spherical screw 4—Rack 5—Washer 6—Holder 7—Pin 8—Conical handle 9—Balls 10—Power cable 11—Perpendicularity adjusting screw

2. Translational motion head

(1) Function For high material removal rate, large discharge gap is used during rough machining. With the same tool electrode, finish machining is only possible for horizontal surface by vertical motion of the tool electrode, because the discharge gap between vertical planes is too large.

If the same tool electrode is expected to perform finish machining for side (vertical) surfaces, the feeding motion in horizontal plane is necessary. The translational motion head is developed to perform finish machining and improve machining accuracy for the side surfaces with the same tool electrode. Figure 7-13 illustrates the advantage of performing EDM by a translational motion head.

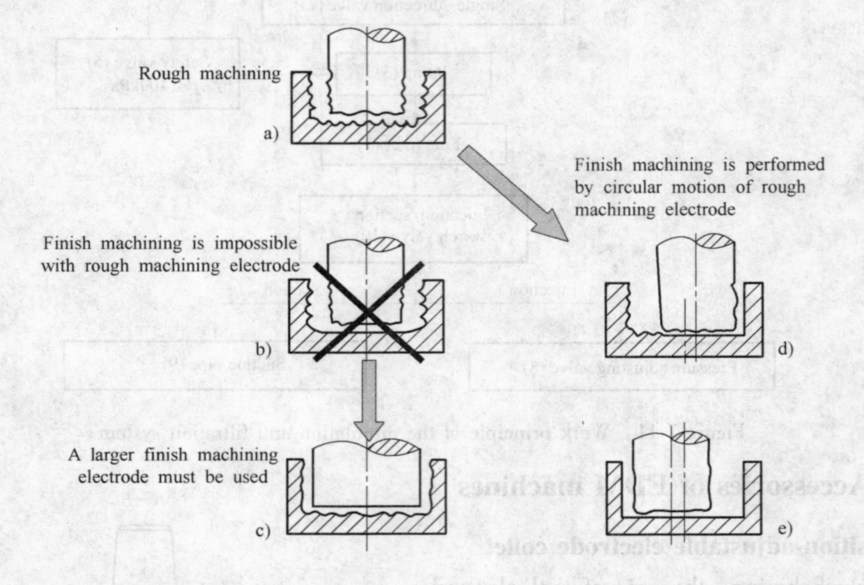

Figure 7-13 Translational motion of tool electrode in EDM

电火花粗加工的放电间隙比精加工的要大。用一个电极进行粗加工时，将工件的大部分余量蚀除后，工件底面和侧壁四周的表面粗糙度值很大，为了将其修光，就得转换成放电间隙小的规准进行修整。对工件底面可通过主轴进给进行修光，而四周侧壁就无法修光。平动头就是为解决修光侧壁和提高其尺寸精度而设计的，它能使工具电极作一定的水平运动，从而实现用同一个电极完成粗、精加工。普通加工与平动加工的比较如图 7-13 所示。

(2) Principle of translational motion The eccentric mechanism converts rotation of the servo motor into translational motion of the tool electrode (see Figure 7-14) about its original position. The motion radius can be adjusted between zero and the maximum to obtain required discharge gap for machining requirements. The motion path forms the contour of finish machining.

平动头利用偏心机构将伺服电动机的旋转运动通过平动轨迹保持机构转化成电极上每一个质点都能围绕其原始位置在水平面内作平面小圆周运动，如图 7-14 所示。这样，许多小圆的外包络线就形成了加工表面。电极的运动半径通过调节可由零逐步扩大，以获得不同的放电间隙，从而达到修光型腔的目的。

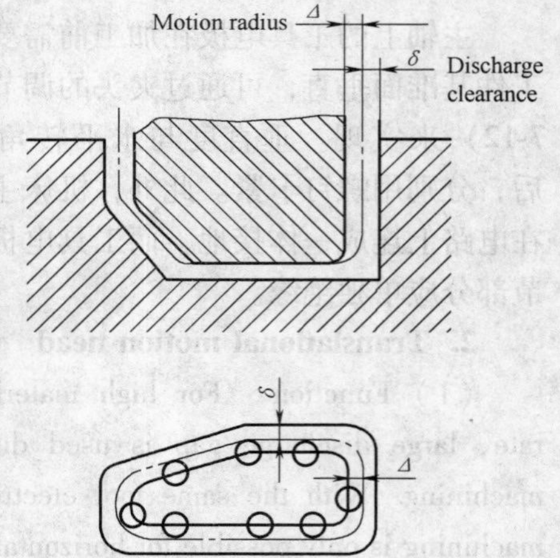

Figure 7-14 Translational motion of tool electrode

(3) Construction of translational motion head The translational motion head is usually consists of an eccentric

mechanism and a motion contour retainer.

① Eccentric mechanism (see Figure 7-15). It is usually a pair of eccentric shaft and sleeve.

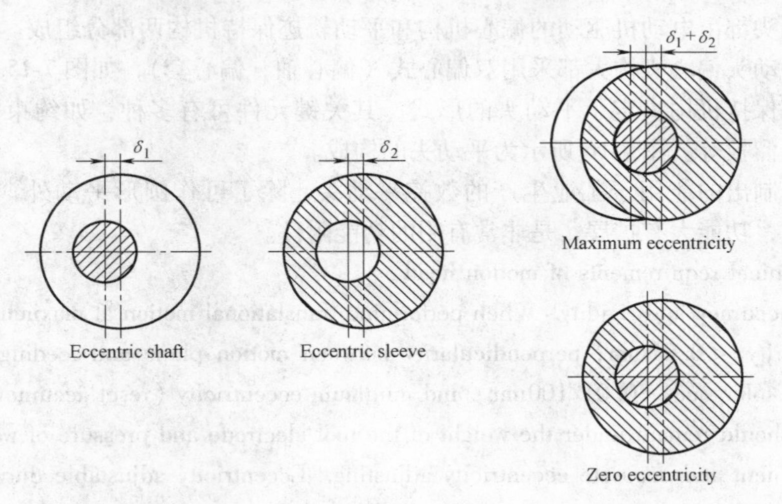

Figure 7-15 Work principle of eccentric mechanism

② Motion contour retainer. A track keeping mechanism actually regulates the motion contour. The functional components can be in many types, e. g. spring strips, four-bar linkage, cross-rolling slider, etc. Figure 7-16 illustrates the construction of a translational motion head.

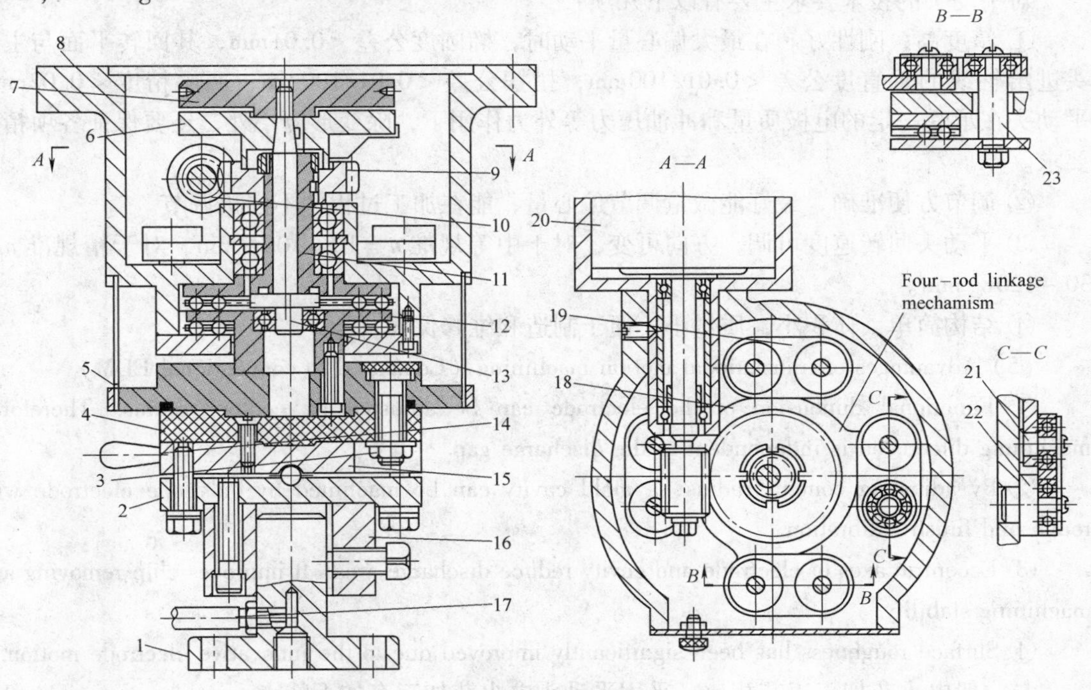

Figure 7-16 Construction of a translational motion head

1—Tool handle 2、5、15—Flange 3、7—Nuts 4—Insulating plate 6—Dial 8—Housing 9—Worm gear 10—Worm shaft 11—Eccentric sleeve 12—Supporter 13—Eccentric shaft 14—Handle 16—Collet 17—Oil pipe 18—Transition plate 19—Chain plate 20—Servo motor 21、22、23—Shaft

Recently, industrial CNC motion heads are available. Besides translational motion, it is able to perform X-motion or cross-motion.

一般平动头都由电动机驱动的偏心机构和平动轨迹保持机构两部分组成。

国产的平动头偏心机构大都采用双偏心式（偏心轴、偏心套），如图 7-15 所示。

平动轨迹保持机构决定了平动头的形式，其关键元件可有多种，如约束弹簧片、四连杆、十字滚动溜板等。图 7-16 所示为平动头的构成。

近几年研制出的能用于工业生产的数控平动头，除了可作圆形平动外，还可作 X 形、十字形平动等，功能大有扩展，是非常有用的功能附件。

(4) Technical requirements of motion head

① High accuracy and rigidity. When performing translational motion at maximum eccentricity, contour circularity <0.01mm; perpendicularity between motion plane and feeding axis <0.01/100mm; swing tolerance <0.01/100mm; and minimum eccentricity (reset accuracy) <0.02mm. The accuracy should remain under the weight of the tool electrode and pressure of work fluid flow.

② Convenient and accurate eccentricity adjusting. Eccentricity adjustable during operation is preferred.

③ Adjustable motion speed for desired surface roughness. For medium calibration, n = 10-100r/min, and for fine calibration n = 30-120r/min.

④ Simple and compact construction for manufacturing and maintenance.

对平动头的技术要求主要有以下几项：

① 精度高，刚性好。在最大偏心量平动时，椭圆度公差 <0.01mm，其回转平面与主轴头进给轴线的垂直度公差 <0.01/100mm，扭摆公差 <0.01/100mm，回零精度 <0.02mm。平动头在承受一定的电极质量和冲油压力等外力作用下，除变形应小外，还要保证各项精度要求。

② 调节方便准确，最好能微量调节偏心量，能在加工过程中不停机调节。

③ 平动头回转速度可调，方向可变，对于中等规准 n = 10 ~ 100r/min，对于精规准 n = 30 ~ 120r/min。

④ 结构简单，体积小，质量小，便于制造和维修保养。

(5) Advantages of translational motion machining (Compared to conventional EDM)

① Functional dimension of the electrode can be adjusted by contour radius. Therefore, machining dimension is independent to the discharge gap.

② By adjusting contour radius, a mold cavity can be machined by only one electrode with rough and finish calibration.

③ Eccentric axes of electrode and cavity reduce discharge area. It improves chip removing and machining stability.

④ Surface roughness has been significantly improved due to the innovative electrode motion.

与一般电火花加工工艺相比，采用平动头电火花加工有如下特点：

① 由于电极的轨迹半径可调，加工尺寸不再受放电间隙的限制。

② 用同一尺寸的工具电极，通过轨迹半径的改变，可以实现由粗至精直接加工出一副型腔。

③ 加工过程中，除了放电区域，工具电极与工件的间隙都大于放电间隙，有利于排除电蚀残物，提高加工稳定性。

④ 工具电极移动方式的改变大大改善了加工面的表面粗糙度。

3. Oilcup

The oilcup (see Figure 7-17) is an important accessory of work fluid circulation system. It has an injection nozzle on side-wall and suction nozzle on the bottom. The main function of oilcup is flushing chips away during EDM. It also drains inflammable gas produced during EDM process. Otherwise, gathered gas may be ignited and explodes to break the relative position between the oilcup and the workpiece thus affecting machining accuracy.

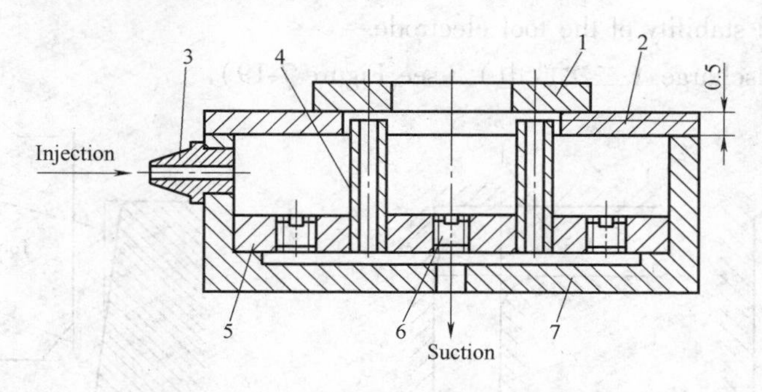

Figure 7-17 Construction of an oilcup

1—Workpiece 2—Cover 3—Connector 4—Extractor 5—Base 6—Oil plug 7—Body

The oilcup should have enough depth, and can perform both work fluid injection and suction. Gas gathering at the top must be avoided.

Good accuracy and rigidity are the other requirements (parallelity between top and bottom surface <0.01mm). In order to prevent oil leakage, sealing must be kept in normal condition. The extracting nozzle can also be placed on side-wall if necessary.

在电火花加工中，油杯是工作液循环中的一个主要附件，其侧壁和底边上开有冲油和抽油孔，如图 7-17 所示，放电加工时，可使电蚀残物及时排出。放电加工过程中工件会分解产生气体，被电火花放电引燃时，将会发生爆炸，造成电极与工件位移，影响被加工工件的尺寸精度。因此，油杯必须有排气的功能。

油杯的技术要求主要是有合适的高度，应具备冲、抽油的条件，但不能在顶部积聚气泡。另外，刚度和精度要符合要求，油杯的两端面平行度应小于 0.01mm，同时密封良好，不可漏油。

7.1.6 Machining quality effectings

1. Contour forming effectings

Similar to conventional machining, the machine tool and the tool electrode mounting errors affect machining accuracy. Besides, some inherent factors may affect the accuracy of EDM:

① Dimension and consistency (一致性) of the discharge gap.

② Shape of the tool electrode (e.g. acute angles and sharp corners) (see Figure 7-18).

The acute angle or sharp corner on the tool electrode can be hardly reproduced onto the workpiece, because the position where an acute angle is to be machined on the workpiece has larger probability of electrical erosion thus to be a fillet. When using a tool electrode with an acute angle to machine a sharp corner, the acute angle can only produce an arc with the center of the vertex of the angle and the radius of the discharge gap because of discharge isometry. Moreover, the acute angle of the tool electrode will soon become a fillet due to the electrical erosion.

The solution is machining by high frequency and narrow width pulses. The method can significantly decrease the fillet radius (down to 0.01mm) thus improving accuracy. It is especially important for machining molds that are used to produce precise gears.

③ Wearing and stability of the tool electrode.

④ Incidental discharge（二次放电）(see Figure 7-19).

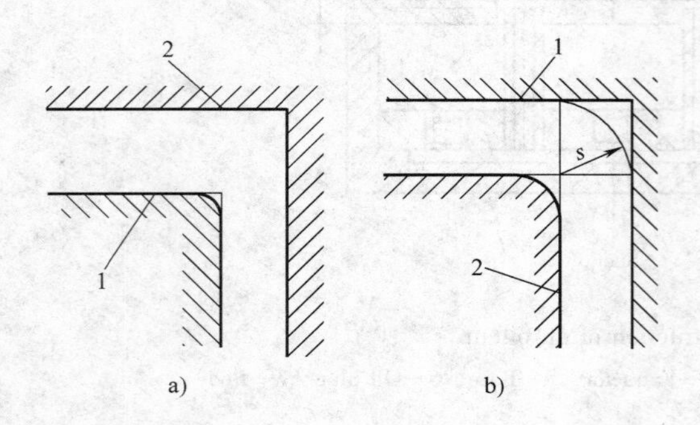

Figure 7-18 EDM for acute angle and sharp corner

a) Acute angle machining b) Sharp corner machining

1—Workpiece 2—Tool electrode

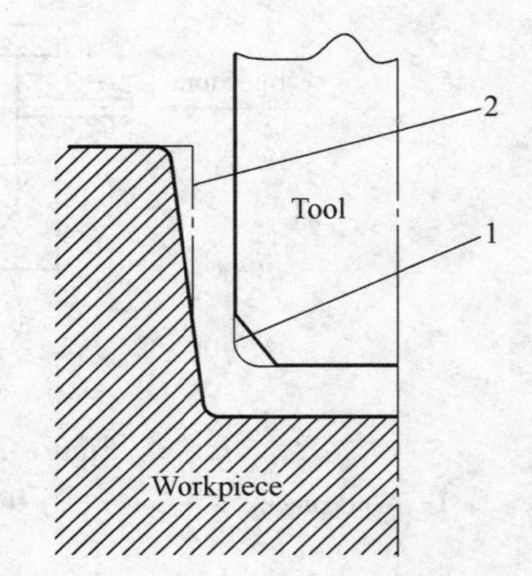

Figure 7-19 Slope due to incidental discharge of EDM

1—Contour of worn electrode 2—Desired contour of workpiece

与通常的机械加工一样，电火花加工中机床本身的误差以及工件和工具电极的定位、安装误差都会影响到加工精度。影响电火花加工工艺的因素主要有以下几项。

① 放电间隙的大小及其一致性。

② 复杂形状的加工表面，特别是棱角部位电场强度分布不均对加工工艺会产生影响，如图 7-18 所示。

③ 工具电极的损耗。

④ “二次放电”导致加工斜度，如图 7-19 所示。

2. Surface roughness effectings

Surface roughness is generally determined by pulse energy, which is defined as:

$$E = t_i \times i_e$$

Where: t_i is the pulse width, and i_e is the peak amperage.

It was found in practice that large discharge area worsens surface roughness, usually, $Ra > 2\mu m$. Material characters of the workpiece also affect surface roughness. Under the same discharge energy, the material that has higher melt point (e.g. hard alloy) has better roughness quality than those have lower melt point (e.g. steel), but machining speed is lower.

影响表面粗糙度的因素主要是单个脉冲能量的大小，即脉宽与峰值电流的乘积。实践中

发现，加工面积越大，可达到的最佳表面粗糙度越差，一般表面粗糙度值 *Ra* 很难低于 2μm。工件材料对加工表面粗糙度也有影响，在相同能量下，对熔点高的材料（如硬质合金）加工的表面粗糙度要比熔点低的材料（如钢）的好，当然，加工速度会相应下降。

Surface roughness of the tool electrode significantly affects machining quality. For example, the surface of the graphite electrode is hard to be smooth, therefore, the surface of the workpiece machined by the graphite electrode has worse roughness.

For EDM, surface roughness conflicts machining speed. To obtain a roughness of *Ra* = 1.25μm, machining speed is less than 1/10 of that for *Ra* = 2.5μm.

Today, it is difficult to obtain a roughness better than *Ra* = 0.32μm, however, with translational motion technique, it can be significantly improved.

精加工时，工具电极的表面粗糙度也会影响加工表面粗糙度。例如石墨电极的表面很难加工光滑，因此用石墨电极加工的表面粗糙度较差。

电火花加工的表面粗糙度和加工速度是矛盾的，例如从 *Ra*2.5μm 提高到 *Ra*1.25μm，加工耗时要增加十多倍。按目前的工艺水平，较大面积的电火花加工要达到优于 *Ra*0.32μm 是比较困难的，但是如果采用平动工艺，加工表面表面粗糙度可大为改善。

3. Surface structure effectings

After EDM, material structure on the workpiece surface has been changed by many factors. For example, carbon penetrating into solidification layer produces micro-cracks, which reduces fatigue strength of the workpiece. Again, high temperature during discharge and fast cooling of the work fluid produce a heat affected layer (> 60HRC). Though high hardness of metamorphic layer improves abrasion-resistance, it also makes grinding and polishing process difficult and time consuming.

Some structure change has advantage, for example, large number of micro-cavities on the surface after EDM forms a layer whose microstructure is like sponge to hold lubricant.

经电火花加工后的工件表面的材料性能有所变化。例如，工作介质和石墨电极的碳元素渗入凝固层而使工件表面有许多微细裂纹，工件疲劳强度下降。再如热影响层位于凝固层和工件基体材料之间，由于受到放电高温影响，材料的金相组织发生了变化，一般硬度可达 60HRC 以上，工件耐磨性和使用寿命都大大提高，但给后续加工工序（研磨、抛光等）增加了困难。

采用回火处理和喷丸处理等有助于降低残留应力，或使残留拉应力转变为压应力，从而提高其耐疲劳性能。

但是，也有一些有益的结构改变，如大量的微孔可吸附润滑油，提高材料的润滑性能。

7.2 Wire EDM machine

7.2.1 Basis of wire EDM

1. Basic principle

As the tool electrode, a fast running metal wire (usually a string made of brass or molybdenum) performs continuous electric erosion to workpieces for desired shapes.

The metal wire connects cathode of high frequency pulse generator while the workpiece connects the anode.

电火花线切割加工的基本原理是利用移动的细金属丝（铜丝或钼丝）作为工具电极（接高频脉冲电源的负极），对工件（接高频脉冲电源的正极）进行脉冲火花放电，切割成形。

By wire travel speed, wire EDM machines can be classified to high speed and low speed series.

The high speed wire EDM (WEDM-HS) is developed by Chinese. Electrode wire performs reciprocate moving of 8-10m/s for fast machining. The wires are reusable and easy to obtain stable discharge. However, high speed vibration and pause during direction shifting affect surface quality of workpieces. Figure 7-20 shows the configuration of a high speed wire EDM.

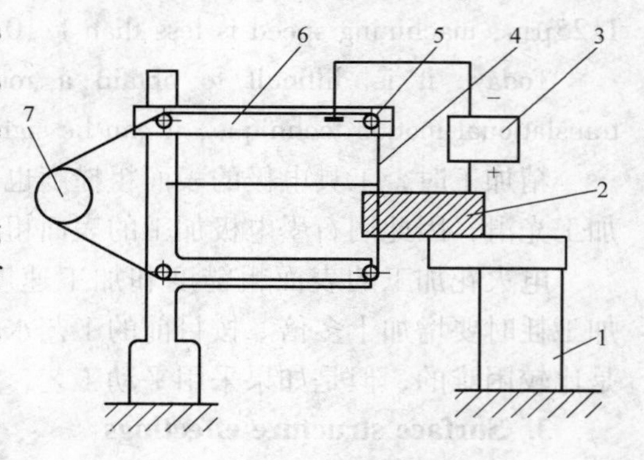

Figure 7-20 Configuration of a high speed wire EDM

1—Insulated base 2—Workpiece 3—Pulse generator 4—Molybdenum wire 5—Guide pulley 6—Support 7—Wire winch

The low speed wire EDM (WEDM-LS) is widely applied in most foreign countries. The electrode wire moves in single direction at < 0.2m/s. After machining, the wire will be rejectes. Stable and smooth machining process improves machining quality.

根据电极丝的运行速度，电火花线切割加工机床通常分为高速走丝和低速走丝两大类：

高速走丝线切割机床是我国生产和使用的主要机种，也是我国独创的电火花线切割加工模式，走丝速度一般为8~10m/s，电极丝可重复使用，加工速度较高，但容易造成电极丝抖动和反向时停顿，使加工质量下降。高速走丝电火花线切割机床布局如图7-20所示。

低速走丝线切割机床的电极丝作低速单向运动，一般走丝速度低于0.2m/s，电极丝放电后不再使用，工作平稳、均匀，抖动小，加工质量较好，但加工速度较低，是国外生产和使用的主要机种。

Wire EDM machine tools have generally three different contour control methods.

① Template contouring. A probe moves along the edge of a profile model to guide the relative moving between the workpiece and cutting wire.

② Photoelectric tracing. The contour of the workpiece has been drawn on a paper. During machining, a photoelectric probe traces the pattern. Decoding circuit converts the pattern into control signals for executors.

③ Program control. Numerical system controls executors according to instructions for machining. Neither profile model nor drawing is needed. Convenience and high accuracy make the control method be widely applied. Today, more than 95% of wire EDM machine tools in the world are numerical controlled.

根据对电极丝运动轨迹的控制形式不同，电火花线切割机床可分为三种：

① 仿形控制。加工前预先制造出与工件形状相同的靠模，加工时把工件毛坯和靠模同时装夹在机床工作台上，在切割过程中电极丝紧紧地贴着靠模边缘移动，从而切割出与靠模形状和精度相同的工件来。

② 光电跟踪控制。加工前先绘制一张零件轮廓图，加工时将图样置于机床的光电跟踪台上，光电头将始终沿着墨线的轨迹运动，再借助于电气、机械的联动，控制机床工作台连同工件相对电极丝作相似形的运动，从而切割出与图样形状相同的工件来。

③ 程序控制。加工前根据工件几何形状参数预先编制好数控加工程序，数控系统按照程序驱动机床自动完成加工。这种方式比前两种控制形式更为方便，加工精度更高。目前95%以上的电火花线切割机床都已数控化。

2. Features of wire EDM

Besides the same features of EDM, the wire EDM has these advantages:

① Metal (brass or molybdenum) wires are used for cutting tool instead of tool electrode. It is the simplest "cutting tool" that significantly reduces preparing time before machining.

② With CAD auto-programmer, 2D contour machining can be conveniently realized.

③ Though wearing exists, fast moving (WEDM-HS) or frequent replacing (WEDM-LS) of the wires minimizes dimension error.

④ Compared to a milling cutter, the wire ($d = \phi 0.025$-$\phi 0.3$mm) is able to cut a workpiece of small dimensions (see Figure 7-21). It also saves material due to lower chip/workpiece mass ratio. Slugs from the wire EDM may be reusable whereas chips from conventional machining are recyclable at best.

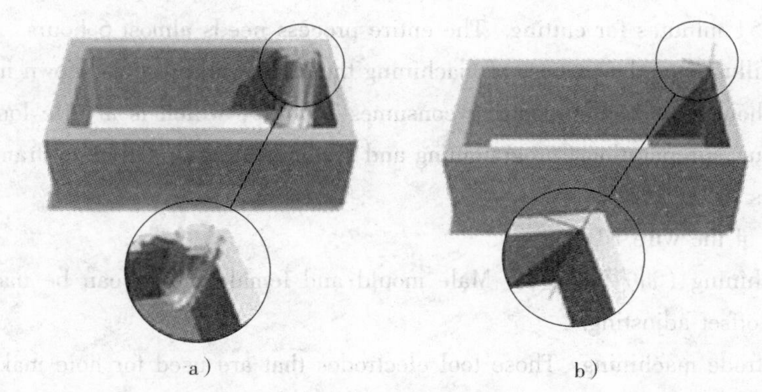

a)　　b)

Figure 7-21 Machining at small radius

a) Round corner machining by milling　b) Sharp corner machining by wire EDM

⑤ With the function of cutting path offset, fitting clearance of moulds can be conveniently adjusted. Even a pair of mould can be cut from the same piece of material.

⑥ Rough and finish machining can be realized by simply adjusting electric pulse parameters.

⑦ Improved structure and manufacturability of workpieces.

Though the wire EDM has many advantages, it can be used only for 2D contour machining, and most tapered surface (with special components and controllers of the machine tool). Stepped holes and blind holes machining are impossible. Moreover, drilling must be performed for wire going through before any hole making.

电火花线切割加工具有电火花加工的共性，此外，还有其专有的工艺特点：

① 不需要特定形状的工具电极，而是采用细金属丝（铜丝或钼丝等）作为工具电极，因此“刀具”极其简单，大大降低了生产准备工时。

② 利用计算机辅助制图自动编程软件，可方便地加工形状复杂的直纹表面。

③ 电极丝在加工中是移动的，不断更新（低速走丝）或快速移动（高速走丝），可以完全或短时间不考虑电极丝损耗对加工精度的影响。

④ 电极丝直径较细（$\phi0.025 \sim \phi0.3$mm），切缝很窄，这样不仅有利于材料的利用，而且适合细小零件的加工。

⑤ 依靠计算机对电极丝轨迹的控制和偏移轨迹的计算，可方便地调整凹凸模具的配合间隙，有可能实现凹、凸模一次加工成型。

⑥ 对于粗、精加工，只需调整电参数即可，操作方便、自动化程度高。

⑦ 有利于改进工件的结构和可加工性。

线切割的应用也有一定的局限性，如加工对象主要是平面形状（除非当机床加上能使电极丝作相应倾斜运动的功能才能实现锥面加工），对于阶梯孔和不通孔还无法加工，当零件无法从周边切入时，工件上需钻穿丝孔。

Here is an example that compares the efficiency of conventional machining and wire EDM machining.

Figure 7-22 illustrates the sixteen steps of machining such a workpiece by mechanical cutting methods. It consumes 4 hours for machining preparation and equipment system set up, and consumes 1 hour 51 minutes for cutting. The entire process needs almost 6 hours.

Figure 7-23 illustrates the process of machining the same workpiece as shown in Figure 7-22 by the wire EDM. Though the contour cutting consumes 2 hours, which is a little longer than that of conventional cutting, preparation (programming and system setting up) time is dramatically reduced to only 15 minutes.

Applications of the wire EDM:

① Dies machining (冲模加工). Male mould and female mould can be machined with the same program by offset adjusting.

② Tool electrode machining. Those tool electrodes that are used for hole making in EDM are usually machined by the wire EDM. The wire EDM is especially suitable for machining workpieces that are made of hard alloys (e. g. copper-tungsten alloy, silver-tungsten alloy) and have small dimensions and complex 2D contours.

③ Prototype machining. The new product is preferred to be made by the wire EDM rather than developing a new mould. The wire EDM is good at making various parts in small quantity, e. g. cams, templates, parts with non-round holes, etc.

线切割加工主要应用于下列场合：

① 加工各种形状的冲模。通过调整偏移量，只需一次编程就可以切割凸模、凸模固定板、凹模及卸料板等。还可以用于加工挤压模、粉末冶金模、弯曲模和塑料模等通常带锥度的模具。

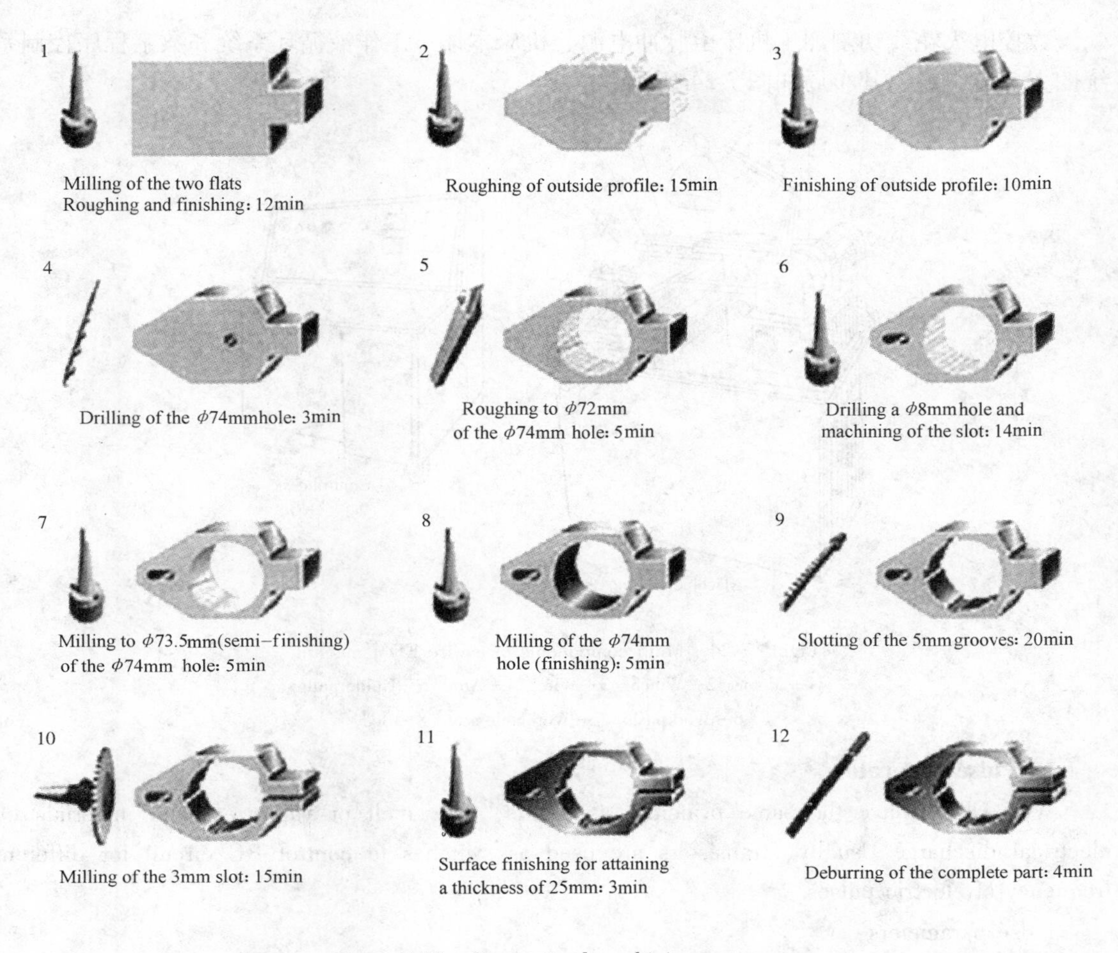

Figure 7-22 Conventional machining process

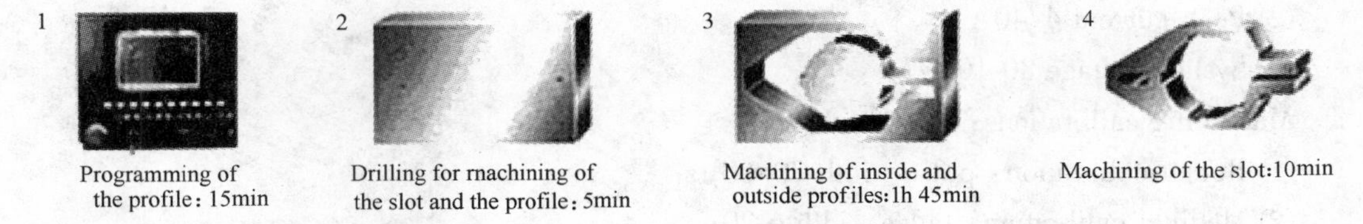

Figure 7-23 EDM process

② 加工电火花加工用的电极。一般穿孔加工用的电极和带锥度型腔加工用的电极，以及铜钨、银钨合金之类的电极材料，都可以用线切割机床来加工。另外也适合加工微细复杂形状的电极。

③ 加工零件。试制新产品时，直接用线切割在坯料上切割零件，不需要另行制造模具，大大缩短了生产周期，降低了成本。

7.2.2 Basic construction of wire EDM machines

Generally, a wire EDM machine consists of four systems, namely, pulse generator, machine tool reality, work fluid circulation system and CNC system, as shown in Figure 7-24.

数控电火花线切割加工机床由脉冲电源、机床本体、工作液循环系统和数字程序控制系统四大部分组成，其布局如图 7-24 所示。

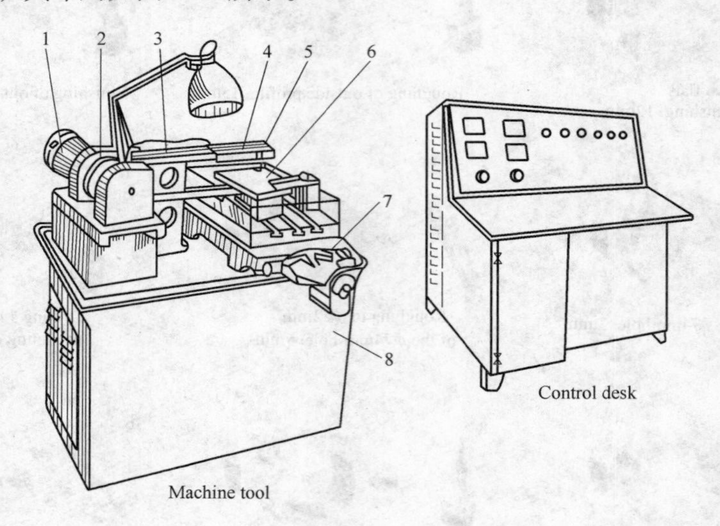

Figure 7-24　Main components of a wire EDM machine

1—Motor　2—Winch　3—Wire　4—Arm　5—Guide pulley

6—Workpiece　7—Worktable seat　8—Bed

1. Pulse generator

Wire EDMs have the same principle as EDMs, i. e. melt or vaporize metal materials by electrical discharge. Usually, transistors are used as switches to control RC circuit for different frequency of electric pulses.

Pulse parameters:

① Pulse interval 10-200μs

② Peak current 4-40A

③ System voltage 80-100V

Machining calibrations:

① Rough calibration: pulse width 20-60μs

② Medium calibration: pulse width 6-20μs

③ Finish calibration: pulse width 2-6μs

电火花线切割加工和电火花成形加工一样，都是利用火花放电对金属工件进行电腐蚀加工。高频脉冲利用晶体管作为开关元件，控制 RC 电路进行精微加工，一般电源的电规准设有粗、中、精三种，以满足不同加工的要求。粗规准一般采用较大的峰值电流，较长的脉冲宽度（20～60μs）；中规准采用的脉冲宽度一般为 6～20μs；精规准采用小的峰值电流，高频率和短的脉冲宽度（2～6μs）。脉冲间隙为 10～200μs，峰值电流为 4～40A，加工电流为 0.2～7A，开路电压为 80～100V。

2. Machine tool reality

The machine tool reality includes a bed, a worktable, a wire storage and driving device, a wire arm, guide pulleys, a tapered surface machining device, a work fluid tank, fixtures and other

accessories.

机床本体由床身、工作台、储丝走丝部件、丝架、导轮部件、锥度切割装置、工作液箱、固定装置及附件等几部分组成。

(1) Bed As the base of a machine tool, the bed supports other components. It is usually cast or welded as a framework. Marble beds are applied on some precise wire EDM machines. The pulse generator and the work fluid tank are usually arranged inside the bed for a compact construction. In some configurations, they locate outside of the bed to avoid heat and vibration affecting machining accuracy.

床身是机床本体的基础，支承和安装坐标工作台、绕丝机构及丝架。一般线切割机床的床身为铸造箱式结构和焊接箱式结构，部分精密线切割机床采用大理石床身。床身内部安置电源和工作液箱。考虑到电源的发热和工作液泵的振动，有的机床将电源和工作液箱搬移出床身外另行安置。

(2) Worktable The worktable is the upper surface of the upper slide that performs independent linear motion in *X*-axis. The lower slide carries the upper slide, and performs linear motion in *Y*-axis. Both the upper and lower the slides are driven by servo motors though ball screw-nut system is to perform feeding motion, as shown in Figure 7-25.

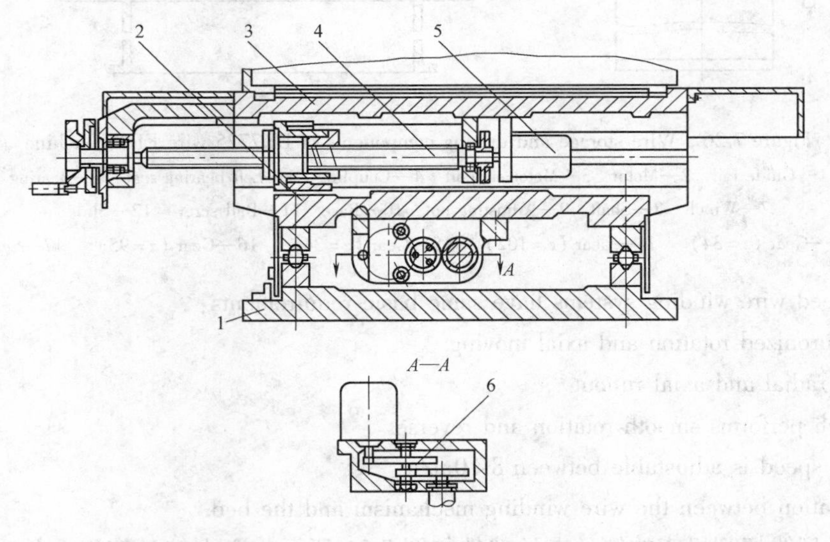

Figure 7-25 Construction of worktable transmission of a wire EDM machine

1—Lower slide 2—Middle slide 3—Upper slide (worktable)

4—Ball screw 5—servo motor 6—Gear transmission

工作台为上滑板的上表面，作独立纵向运动（*X* 坐标）。上滑板安装在下滑板上，由下滑板带动进行横向运动（*Y* 坐标）。工作台的移动由步进电动机带动滚珠丝杠副实现，如图 7-25 所示。

3. Wire storage and driving devices

Figure 7-26 illustrates the components of wire storage and driving devices of DK7725 wire EDM machine. The motor (2) drives the winch (7) at 1, 400r/min through the coupling (4) to wind the wire. The right end of the shaft (8) drives the gear (13), and then drives the ball screw (11)

through gears (14, 15 and 16). The gearings realize a synchronized axial moving of the slide (12) with all the components carried (including the winch) to avoid superposition of the wire. The axial travel distance of the winch during each revolution must be slightly larger than the diameter of the wire in use. For DK7725, the slide (12) carries the winch (7) to perform axial movement of 0.275mm during each revolution of the winch. Therefore, those wires that have diameters of less than ϕ0.25mm can be used in the system.

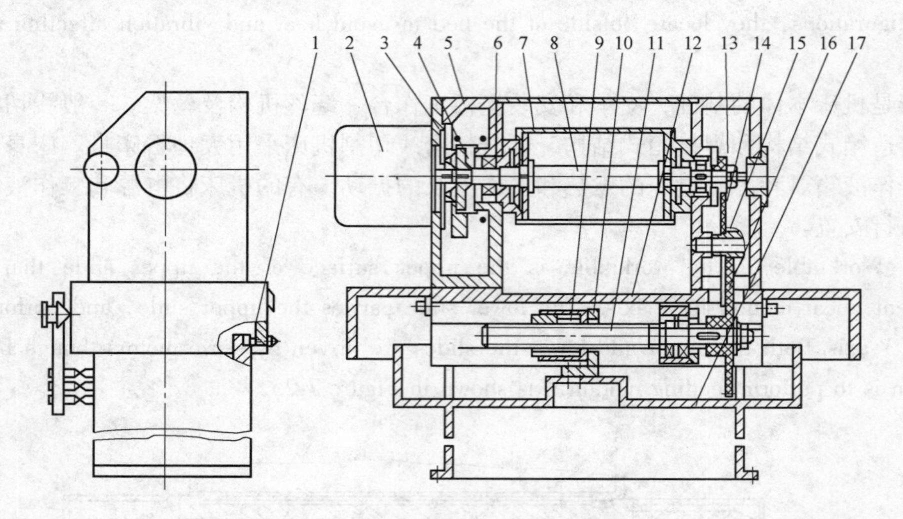

Figure 7-26 Wire storage and driving components of DK7725 wire EDM machine

1—Guide rail 2—Motor 3—Motor support 4—Coupling 5—Left bearing seat 6—Bearing 7—Winch 8—Shaft 9—Adjusting nut 10—Base 11—Ball screw 12—Slide 13—Gear ($z=34$) 14—Gear ($z=102$) 15—Gear ($z=34$) 16—Gear ($z=95$) 17—Seat

High speed wire winding systems have same basic requirements:

① Synchronized rotation and axial moving.

② Less radial and axial runout.

③ Winch performs smooth rotation and reverse.

④ Wire speed is adjustable between 8-10m/s.

⑤ Insulation between the wire winding mechanism and the bed.

DK7725 型线切割机床的储丝走丝部件如图 7-26 所示，它由储丝筒组合件、上下滑板、齿轮副、丝杠副、换向装置和绝缘件等部分组成。储丝筒 7 由电动机 2 通过联轴器 4 带动，以 1400r/min 的转速正、反向转动。储丝筒另一端通过 3 对齿轮副减速后带动丝杠 11。机构中使用的齿轮、丝杠副等元件决定了储丝筒和滑板之间的综合传动比，储丝筒每旋转一周，滑板移动 0.275mm。因此，该储丝筒适用于 ϕ0.25mm 以下的钼丝。

高速绕丝机构的转动和轴向移动必须协调，运动时径向和轴向的窜动量要小。丝筒的转动和换向过程应尽量平稳。电极丝的走丝速度在 8 ~ 10m/s 的范围内可调。绕丝走丝机构与床身之间必须绝缘良好。

The low speed wireEDM machine has a different wire winding system, which is shown in Figure 7-27. The wire storage (1) carries 1-3kg of wire for machining. The driving wheel (7) pulls the wire

and rotates other components through the wire. By adjusting resistance torque, the braking wheel (3) controls the wire tension. The wheel (8) keeps the wire winding on the wheel (7) in order. Wire guiders (14, 16) eliminate the effect of wire vibration, and precisely ensure the relative position between the wire and the workpiece. The wire break detectors (4, 12) will send a signal to the control system to stop the machine in case of wire break.

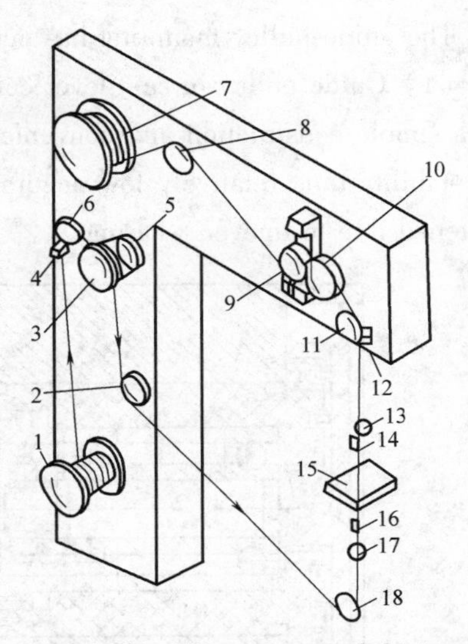

Figure 7-27 Wire storage and driving system of low speed wire EDM machine

1—Wire storage 2、5、6、11、18—Pulley 3—Brale wheel 4、12—Wire break detector 7—Driving wheel 8—Wire aligning wheel 9—Tension wheel 10—Pulley 13、17—Poles 14、16—Wire guiders 15—Workpiece

4. Wire arms and guiding pulleys

The wire arms carry the guiding pulleys to keep a certain angle (usually vertical) between the working segment of the wire and the worktable.

丝架与走丝机构组成了电极丝的运动系统。丝架的主要是在电极丝按给定丝速运动时，对电极丝起支承作用，并使电极丝的工作部分与工作台平面保持一定的几何角度。

Figure 7-28 shows a pair of wire arms (static lower arm) of a wire EDM machine. Screw-nut transmission (6) drives the upper arm (7) to perform vertical moving to optimize the distance between the upper guiding pulley and the workpiece for accuracy.

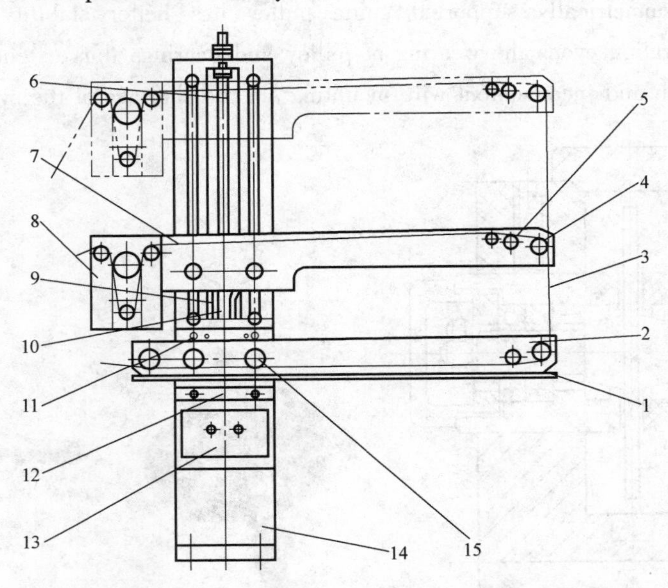

Figure 7-28 A pair of wire arms of a wire EDM machine

1—Gap 2—Lower arm 3—Wire 4—Guiding pulley 5—Conductive pulleys 6—Screw 7—Upper arm 8—Wire tension 9—Electrical cable 10—Coolant pipe 11—Positioner 12—Positioning seat 13—Coolant valve panel 14—Column 15—Adjusting screw

The guide pulley mounting has generally two configurations：

（1）Guide pulley in cantilever configuration（see Figure 7-29） The Cantilever configuration has a simple construction and convenient wire mounting. However, the unsymmetrical structure has less stability thus relatively low accuracy of machining. Duty life of pulleys and bearings are also shortened due to uneven wearings.

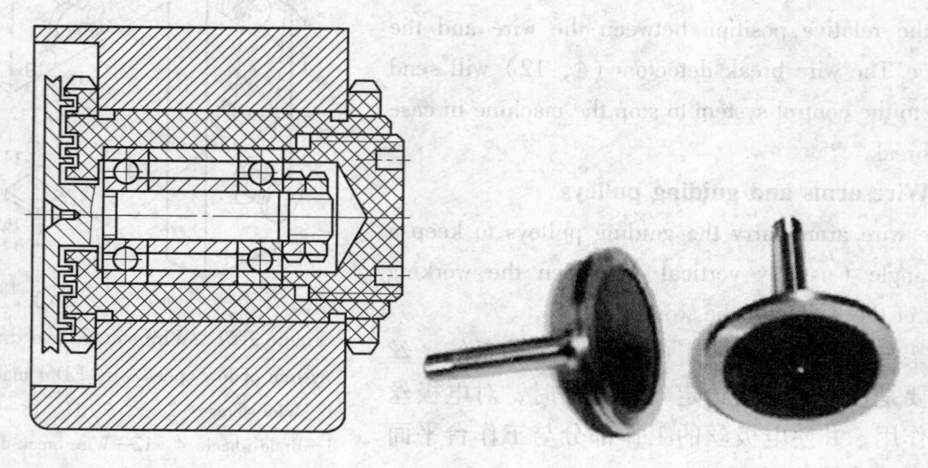

Figure 7-29 Guide pulley in cantilever configuration

悬臂支承导轮（如图 7-29）结构简单、上丝方便，但因悬臂支承，张紧的电极丝运动时稳定性较差，因此加工精度较低，同时导轮和轴承因磨损不均匀而使用寿命缩短。

（2）Symmetrically supported guiding pulley（see Figure 7-30） Compared to a cantilever configuration, a symmetrically supported guide pulley has better stability and rigidity. The symmetrical configuration evens the wearing of pulley and bearings thus extending their duty life. Complex construction and inconvenient wire mounting are the shortage of the system.

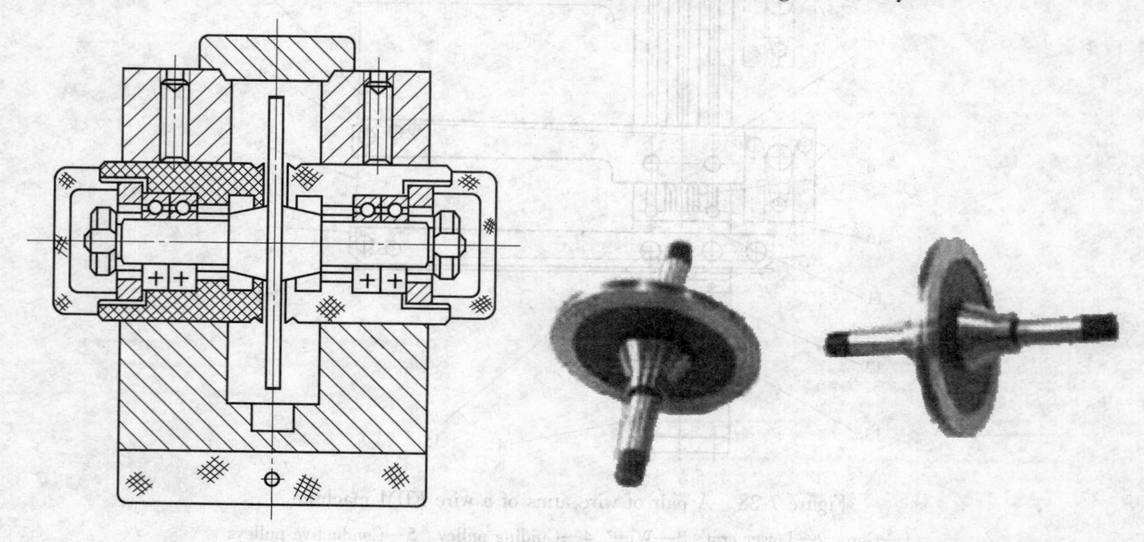

Figure 7-30 Symmetrically supported guiding pulley

双边支承导轮结构如图 7-30 所示。其中导轮居中，两端用轴承支承，结构较复杂，上丝较麻烦，但这种结构的运动稳定性较好，刚度较高，不易发生变形及跳动。

As a key component of a wire EDM machine, the guide pulley determines machining quality. It must meet these requirements:

① The V-groove on the guide pulley should be accurately machined. Diameter at the bottom of the groove must be slightly smaller than that of the wire to keep the wire in track and eliminates wire slipping.

② A light guide pulley is preferred to reduce abrasion between the pulley and wire during direction changing. Work surface of the groove must be wearing-resistant.

③ A guide pulley should rotate freely without any radial and axial runout.

④ Effective sealing must be considered to provide a normal work environment for the bearings.

导轮是线切割机床的关键零件，影响到切割的质量，因此有以下技术要求：

① 导轮 V 形槽面应有较高的精度，V 形槽底的圆弧半径应略小于所用电极丝的半径，以保证电极丝不会在导轮槽内产生轴向移动。

② 在满足强度要求的前提下，应尽可能使导轮轻量化，以减小其惯性，迅速随电极丝换向，减少电极丝与导轮槽间的滑动摩擦。对导轮槽工作面进行表面硬化处理，提高其耐磨性。

③ 装配好的导轮转动应轻便灵活，尽量减小导轮工作时的轴向窜动和径向跳动，确保电极丝平稳通过。

④ 导轮轴支承部分应进行有效密封，保证轴承的正常工作条件。

5. Conical surface machining mechanisms

Though a wire EDM machine is mainly used for 2D contour cutting, conical surfaces machining function is necessary. The function is realized by wire arms' motion. Figure 7-31 shows the principles of three conical surface machining mechanisms.

Type a—One of the wire arms performs translational motion (平动) in *X-Y* plane to incline the wire for tapered machining. However, large inclination results in severe abrasion of guide pulleys, and increases risk of wire break.

Type b—Both the upper wire arm and the lower wire arm perform translational motion about the middle point of the effective wire length. The inclination angle is also limited due to poor engagement condition between the wire and the V-groove on the pulleys.

Type c—Both the upper wire arm and the lower wire arm perform translational motion parallel to *Y*-axis and swing in *X-Z* plane. In this mechanism, the engagement condition between the wire and the V-groove on the pulleys remains its normal status. The conical degree can be up to 1.5°.

图 7-31a 所示为上（或下）丝臂在 *X-Y* 平面内作平动。此法加工锥度不宜过大，否则钼丝易拉断，导轮易磨损。

图 7-31b 所示为上、下丝臂同时绕中心平移。此法加工锥度也不宜过大。

图 7-31c 所示为上、下丝臂分别沿导轮径向平动和轴向摆动的方法，此法加工锥度不影响导轮磨损，最大切割锥度可达 1.5°。

Figure 7-32 shows the construction of the wire arm moving and swing mechanism.

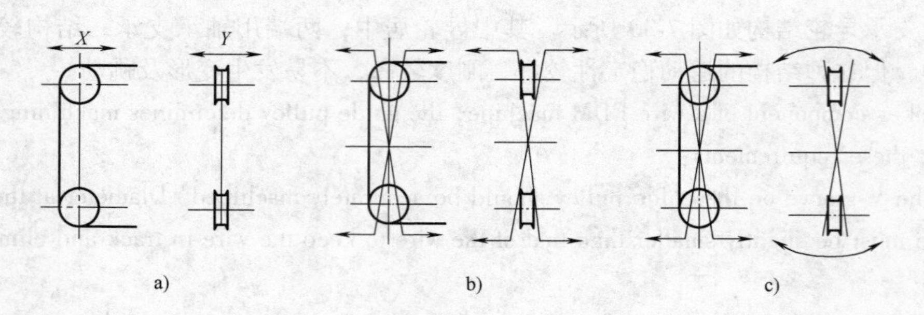

Figure 7-31　Principles of conical surface machining

① Swing motion. The motor (1) drives the screw shaft (4) through the gears (2, 3). The nut (5) then performs axial motion with the slide (6) along the screw shaft (4). Since the pivot (7) is fixed on the base (say, bed), the relative position between the screw shaft (4) with all the components fixed together and the pivot (7) is changed. Then the entire system including the wire arms swings about the axle (16).

② Translational motion. The motor (14) drives the screw shaft (13) thus making the nut (12) perform axial motion with the slide (11). As shown in Figure 7-32, the guide pin moving in *Y*-direction with the slide (11) results in moving in *X*-direction of the guide plate (9) with the upper arm (8).

6. Work fluid circulation system

Though work fluid applied to the wire EDM has different compositions and features, any type of work fluid should have following characteristics:

① Forming an insulative layer between the wire electrode and the workpiece to avoid long time arc discharge.

② Flushing chips away for normal machining.

③ Cooling the workpiece to prevent thermal distortion.

Also, work fluid that is environmentally friendly and safe to human beings is preferred today.

在线切割加工中，工作液对加工工艺指标的影响很大，其成分与特点也各不相同。但不管哪种工作液，都应具有一定的绝缘性能、较好的洗涤性能、较好的冷却性能，且要求对环境无污染，对人体无危害。

Work fluid is injected to machining area through a nozzle. Fluctuation of work fluid flow from a plane nozzle (see Figure 7-33, left) may cause wire vibrating. Circular nozzle (see Figure 7-33, right) significantly reduces wire vibration since the radial force of work fluid flow applied on the wire is symmetrical. Therefore, circular nozzle is strongly recommended for large effective wire length, especially where stability is vital.

工作液通过喷嘴喷射到工件与电极丝之间。图 7-33 左图所示的喷嘴，工作液液流径向冲击电极丝，加上工作液本身的流量波动，容易使电极丝产生振动。当丝架的跨距较大，且工作液产生的冲击力和振动会影响到工件的精度时，建议尽可能使用图 7-33 右图所示的环形喷嘴结构，它可避免工作液对电极丝的直接冲击，且工作液顺着电极丝的运动方向射入，有利于保持电极丝稳定。

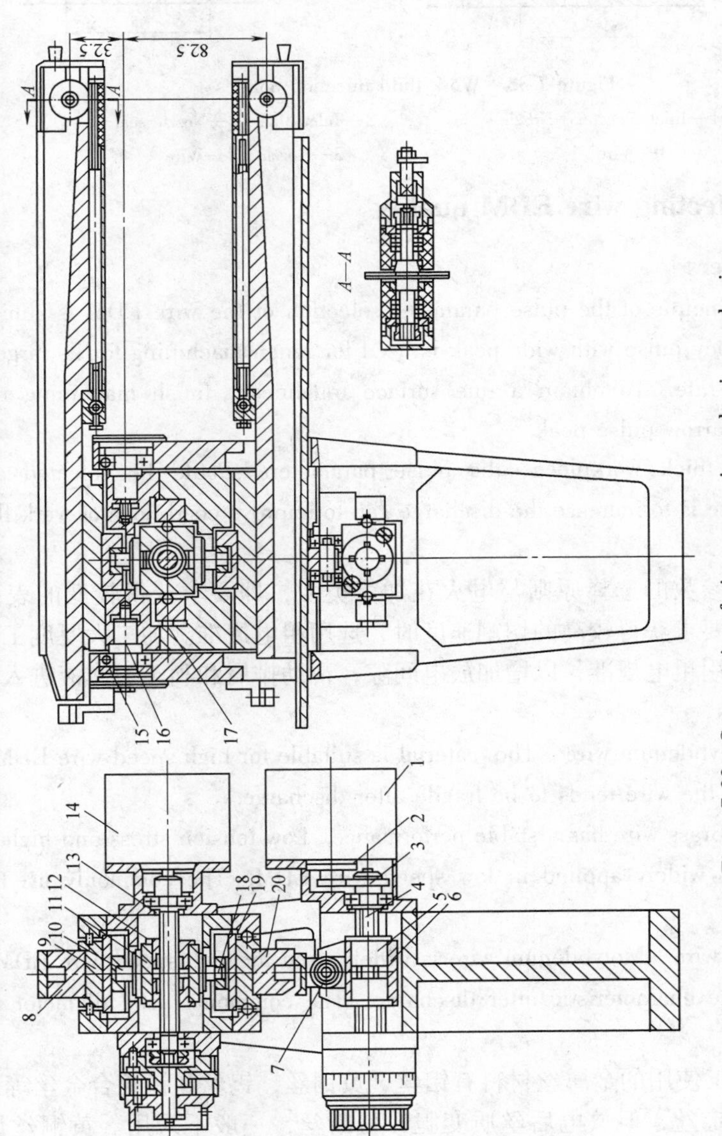

Figure 7-32 Construction of wire arm moving and swing mechanism

1、14—Servo motors 2、3—Gears 4、13—Screw shafts 5、12—Nuts 6、11—Slides 7—Pivot
8、20—Wire arms 9、19—Guide plates 10、18—Guide pins 15—Bearing 16—Axle 17—Base

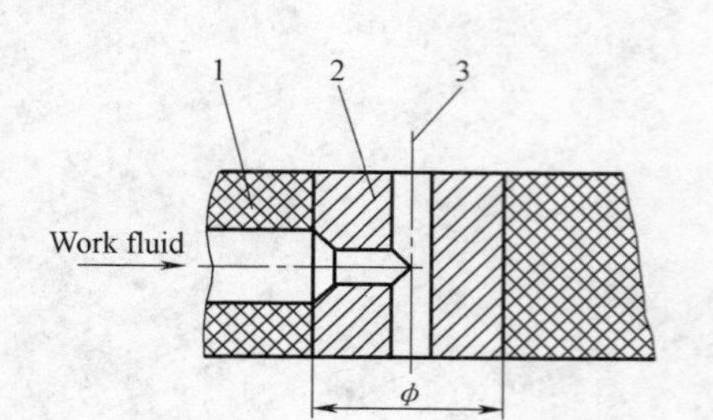

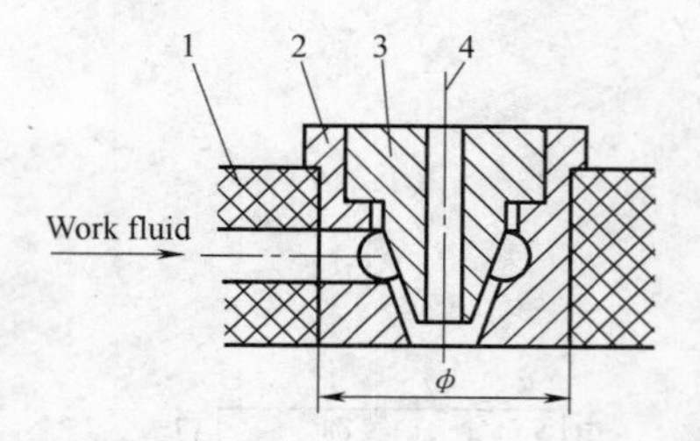

Figure 7-33　Work fluid injection nozzles

1—Inlet pipe　2—Nozzle　3—Wire　　1—Inlet pipe　2—Nozzle seat　3—Nozzle　4—Wire

7.2.3　Factors affecting wire EDM quality

1. Pulse parameters

Generally, the principle of the pulse parameter selection of the wire EDM is similar to that of the EDM. Low frequency pulse with wide peak is used for rough machining for its large energy thus high material removal rate. To obtain a fine surface roughness, finish machining requires high frequency pulse with narrow pulse peak.

For machining a thick workpiece, the pulse parameters should be selected as for rough machining. The purpose is to increase the discharge gap to improve conditions of work fluid flow and chip removal.

线切割加工脉冲参数的选择原则与电火花加工类似，即要求获得较低的表面粗糙度值时，选用精电规准；要求获得较高的切割速度时，选用粗电规准。加工大厚度工件时，为了改善排屑条件，应选用粗电规准，以增加放电间隙，帮助排屑和利于工作液进入加工区。

2. Wire materials

(1) Tungsten-molybdenum wire　The material is suitable for high speed wire EDM. However, it is seldom used since the wire tends to be fragile after discharge.

(2) Brass wire　Brass wire has a stable performance. Low tension stress and high wearing rate are the shortages. It is widely applied in low speed wire EDM. The commonly used diameter is ϕ0.12-0.3mm.

(3) Molybdenum wire　Molybdenum wire is widely used for high speed wire EDM for its high tension stress and stable characteristic after discharge. The commonly used diameter is ϕ0.08 ~ 0.2mm.

电火花线切割加工使用的金属丝材料有钼丝、黄铜丝、钨丝和钨钼合金丝等。钨丝和钨钼合金丝适合于高速走丝，但放电后丝质变脆，易断丝，一般不采用。黄铜丝加工稳定性好，但抗拉强度低、损耗大，通常用于低速走丝，直径一般选用ϕ0.12 ~ ϕ0.3mm。虽然钼丝的加工速度不如前几种，但抗拉强度高，不易变脆，断丝少，因此在实际中广泛采用钼丝作快速走丝的电极丝，其直径一般选用ϕ0.08 ~ ϕ0.2mm。

High speed wire travel improves machining efficiency. Also, work fluid is likely to be brought by the wire at high speed to flush chips way to obtain a stable discharge process. For high speed wire EDM, wire usually travels at 6-12m/s to prevent the wire from vibration and break.

走丝速度影响加工速度。走丝速度提高，加工速度也提高。高速运动的金属丝能将工作液带入厚度较大的工件放电间隙中，有利于电蚀残物的排出和放电加工的稳定。但走丝速度过高，将引起机械振动，易造成断丝。快速走丝线切割机床走丝速度一般为 6～12m/s。

3. Effects of feedrate

For EDM machines, feedrate affects surface quality. Over-feeding may cause short circuit, and decrease the actual electrical erosion rate. It also leaves sear spots on machined surface of the workpiece.

Under-feeding may cause frequent open circuit due to large gap. In this case, EDM does not take place at most time, and the EDM process is not continous. Under-feeding results in low machining speed, coarse surface and wire break.

Therefore, feedrate should be carefully determined based on the linear rate of the electrical erosion of the workpiece.

电火花线切割进给速度对表面粗糙度的影响较大。进给速度太快，超过工件的蚀除速度，会出现频繁的短路现象，实际进给量小，加工表面有过烧现象。进给速度太慢，滞后于工件的蚀除速度，极间将偏于开路。这两种情况都不利于切割加工，影响加工速度。因此，为了获得表面粗糙度值小、精度高的加工效果，进给速度要维持在接近工件被蚀除的线速度。

Glossary

acute	[əˈkjuːt]	锐角的
anode	[ˈænəud]	（电）正极；阳极
cathode	[ˈkæθəud]	（电）负极；阴极
cavity	[ˈkæviti]	（模具的）腔体
collet	[ˈkɔlit]	筒夹；弹性夹套
discharge	[disˈtʃɑːdʒ]	放电
EDM		放电加工机床（electrical discharge machine）
electrode	[iˈlektrəud]	电极
erosion	[iˈrəuʒən]	侵蚀；腐蚀
filtration	[filˈtreiʃən]	过滤
finish machining		精加工
handwheel	[ˈhændhwiːl]	手轮
hazardous	[ˈhæzədəs]	有危险的
incidental	[ˌinsiˈdentl]	附带发生的；避免不了的
maintenance	[ˈmeintinəns]	维护；保养
molybdenum	[məˈlibdinəm]	钼
perpendicular	[ˌpəːpənˈdikjulə]	垂直的；正交的
pulley	[ˈpuli]	滑轮；带轮
recovery	[riˈkʌvəri]	恢复

rough machining		粗加工
steep	[stiːp]	陡峭的
suction	[ˈsʌkʃən]	吸入；抽吸
superposition	[ˌsuːpəpəˈziʃən]	重叠
symmetrically	[səˈmetrikli]	对称地
tungsten	[ˈtʌŋstən]	钨
unfavorably	[ʌnˈfeivərəbli]	不适宜地
vice versa	[vais] [vəːzə]	反之亦然

Exercises

1. Compared to conventional machining process, what are the advantages and shortages of EDM?

2. If use a cylindrical tool electrode to make a hole in a workpiece, the hole is usually conical. Why does it occur?

3. How to select pulse frequency and amperage for rough and finish machining?

4. Why the wire in winch of a wire EDM never superposes?

5. Wire EDM can machine conical surfaces with the mechanism shown in Figure 7-32. Explain the work principle.

References

[1] Patrick Hood-Daniel, James Floyd Kelly. Build Your Own CNC Machine [M]. Barkley: Apress L P, 2009.
[2] Geoff Williams. CNC Robotics [M]. New York: McGraw-Hill/TAB Electronics, 2007.
[3] Zhou Lan, Chang Xiaojun. Modern CNC Machining Facilities [M]. Beijing: China Machine Press, 2007.
[4] Zhang Bolin. High Speed Cutting Technology [M]. Beijing: China Machine Press, 2002.
[5] Liu Jinchun, Zhao Jiaqi, Zhao Wansheng. Special Machining [M]. 4th ed. Beijing: China Machine Press, 2004.
[6] Tang Caiping. CNC English [M]. Beijing: Publishing House of Electronics Industry, 2004.